ADDITIONAL PRAISE FOR *HOLY HATRED*:

"This work is a thorough treatment of an immense topic. So much has been written about Christian antisemitism, and about the Holocaust, that general readers can sometimes lose sight of the fact that an intimate causal relationship existed between the two. All that had gone before, it seemed, allowed the Holocaust to happen—indeed, it might be said that the Holocaust was in fact the culmination of the previous twenty centuries of Christian animosity towards Jews, a hostility that only intensified with each passing year. Robert Michael has crafted a timely study which deserves to be read by specialist and non-specialist alike, the better to understand the dynamics of Christian Jew-hatred from the time of the early Church fathers to Hitler and beyond. Michael is to be congratulated on bringing light to an area that has in the past all too frequently been darkened by the clouds of superstition, bigotry, and denial. This work is, in short, a praiseworthy achievement—a model of its kind."—Paul R. Bartrop, Honorary Research Fellow, The Faculty of Arts, Deakin University, and Head of History, Bialik College

"*Holy Hatred* is a masterful, beautifully written study of how Christianity and the churches shaped and sustained a lethal antisemitism for almost two millennia. With a full command of both primary and secondary sources, Robert Michael exposes the extent and continuity of Christian racial and theological antisemitism from the earliest foundations of the Church through the Holocaust. Michael powerfully delineates nearly two millennia of the obsessive Christian teachings that dehumanized Jews and the behavior [Christians] justified, including the numerous calls to rid the world of Jews and Judaism. The author shows the critical importance of Christian antisemitism to German and Austrian Nazis at all levels and to the vast populations that sustained them. In unforgettable, even shocking, portraits, Michael reveals the extent to which Christian antisemitism shaped the views of leading German literary figures, composers, philosophers, and political thinkers from left to right. A brilliant work, the work of a true master, a powerful and courageous study, unlike that of so many

who have written on the subject of antisemitism, who have chosen to appease Christians and Christianity. Future studies of the Holocaust or of antisemitism will have to address Michael's work. This is a unique and extremely important book!"
—*Eunice G. Pollack*, University of North Texas

"Following in the footsteps of Poliakov and Flannery, this book offers a powerful description of Christianity's intimate involvement with Judeophobia and anti-Semitism from the gospels forward. In the later chapters, which focus on Central Europe in the modern period, the author pulls no punches in describing the religious traditions and historical precedents upon which the Nazis drew."—*Peter J. Haas*, Abba Hillel Silver Professor of Jewish Studies, Chair, Department of Religious Studies, and Director, The Samuel Rosenthal Center for Judaic Studies, Case Western Reserve University

"If anyone still remains ignorant of the Christian origins of anti-Semitism—ancient and modern—and its contribution to the Holocaust, this book will remedy that bliss: clearly, and in comprehensive detail. Organized, institutionalized defamation of Jews and Judaism is shown to derive from the Christian Bible and its sainted interpreters, while incitement to murder Jews begins with the Church Fathers, Popes, and holy men and contaminates the Christian and Secular West. A history of Christianity's crusading sins against Jews, Judaism, and God, in thought, word and deed."—*Richard Elliott Sherwin*, Professor, Bar-Ilan University

Holy Hatred: Christianity, Antisemitism, and the Holocaust

by
Robert Michael

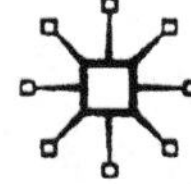

First published in 2006 by
PALGRAVE MACMILLAN™
175 Fifth Avenue, New York, N.Y. 10010 and
Houndmills, Basingstoke, Hampshire, England RG21 6XS
Companies and representatives throughout the world.

PALGRAVE MACMILLAN is the global academic imprint of the Palgrave Macmillan division of St. Martin's Press, LLC and of Palgrave Macmillan Ltd. Macmillan® is a registered trademark in the United States, United Kingdom and other countries. Palgrave is a registered trademark in the European Union and other countries.

ISBN-13: 978–1–4039–7471–6 hardback
ISBN-10: 1–4039–7471–3 hardback
ISBN-13: 978–1–4039–7472–3 paperback
ISBN-10: 1–4039–7472–1 paperback

Library of Congress Cataloging-in-Publication Data

Michael, Robert, 1936–
Holy hatred : Christianity, antisemitism, and the Holocaust / Robert Michael.
p. cm.
Includes bibliographical references and index.
ISBN 1–4039–7471–3 (hardcover : alk. paper)—
ISBN 1–4039–7472–1 (pbk. : alk. paper)
1. Christianity and antisemitism—History. 2. Judaism—Relations—Christianity. 3. Christianity and other religions—Judaism. 4. Holocaust, Jewish (1939–1945)—Causes. I. Title.

BM535.M52 2006
261.2′6—dc22 2006043778

A catalogue record for this book is available from the British Library.

Design by Newgen Imaging Systems (P) Ltd., Chennai, India.

First edition: October 2006

10 9 8 7 6 5 4 3 2 1

Printed in the United States of America.
Transferred to Digital Printing 2007

This book is dedicated to all those people whom I love and who bless my life. In particular, my wife Susan, and my children Stephanie, Andrew, and Carolyn.

The unexamined life is not worth living.

—Plato, The Apology 38a.

Help others, be compassionate, and, at the least, do no harm.

—The Buddha

Do justice, love kindness, and walk humbly with God.

—Micah 6:8

If I am not for myself, then who will be for me? If I am only for myself, what am I? If not now, when?

—Rabbi Hillel, the Elder

Love the Lord your God with all your heart and with all your soul and with all your mind. This is the first and greatest commandment. And the second is like it: Love your neighbor as yourself. All the Law and the Prophets hang on these two commandments.

—Jesus of Nazareth, Matthew 22

Without memory there can be no redemption.

—Ba'al Shem Tov

Courage is a person who keeps on coming on. Slow him down, maybe, but you can't never defeat him.

—Leander McNelly, Captain, Texas Rangers

Be a human being first, a Christian second.

—Nikolai Grundtvig, Danish theologian

The Christians say they love him, but I think they hate him without knowing it. So they take the cross by the other end and make a sword out of it and strike us with it! . . . They take the cross and they turn it around, they turn it around, my God.

—André Schwarz-Bart, The Last of the Just

Contents

Acknowledgments

I would like to thank the following. The libraries without whom this book would have been impossible: University of Massachusetts Dartmouth, Brandeis University, Brown University, Harvard University, Yale University, Ringling School of Art & Design. My colleagues at the University of Massachusetts Dartmouth and at the Humanities Network headquartered at Michigan State University. My teachers at Boston University, Columbia University, Brooklyn College, and the University of Connecticut. The Commonwealth of Massachusetts that supported research grants to me as a faculty member in the commonwealth's higher education structure. In particular, this book would not have been possible without the guidance of Professor Peter A. Bertocci, late of the Philosophy Department of Boston University, who, I am sure, continues to inspire students from a higher mountain.

Chapter 1

Christianity, Antisemitism, and the Holocaust

In the last analysis, antisemitism is a religious problem.

—Jacques Maudaule

Without a thorough knowledge of history, men and women can neither fully understand themselves nor make wise choices for their present and future, for all human beings have been shaped by the values and institutions they have inherited from the past. Of all these values and institutions, those of religion have been primary in their influence on people, who, at their core, are religious beings, *homini religiosi.* Even those not consciously involved with the momentous theological issues of life and death, good and evil, right and wrong are still emotionally caught up in them and often base their most important decisions on ideas, values, and attitudes handed down to them, in the case of the Western world, by the Churches and their theologians.

Of all historical events, the Holocaust seems the most unfathomable. Scholars, especially, understand the inadequacy of historical explanation. And yet, just as historians try to explain the decline and fall of Rome or the causes of the First World War, they struggle to explain the inexplicable, measure the depths of the unfathomable, understand why the Holocaust happened.

Christianity's precise influence on the Holocaust cannot be determined and the Christian churches did not themselves perpetrate the Final Solution. But two millennia of Christian ideas and prejudices, their impact on Christians' behavior, appear to be the major basis of antisemitism and of the apex of antisemitism, the Holocaust.

In the earliest centuries of the Christian era, preexisting pagan antagonism toward Jews (about a quarter of pagan writers were hostile to the Jews because pagans could not understand, on the one hand, Jewish monotheism, and on the other, the Jewish Sabbath, circumcision, and kosher foods) was replaced by the conviction that Jews, all Jews, were forever responsible for murdering God. And so the Jewish people were abhorrent and any injustice done to them, short of murder, according to Augustine, was justified—and even murder was sometimes justified. The "deicidal" Jews became the archetypal evildoers in Christian societies. This anti-Jewish attitude became a permanent element in the fundamental identity of Western Christian civilization. Christian writers transformed Jewish virtues into vices, and transvaluated Jewish values into sins. They called "evil good and good evil . . . everything was completely turned upside-down."[1] This theology assumed that the Christian Church, the "new Israel"—ordained and sanctioned by God—succeeded the cursed and rejected old Israel morally, historically, and metaphysically. This ideology, often termed triumphalism, or *theologia gloriae*, considered Jews an inherently evil people who, long before the birth of Jesus of Nazareth, slaughtered their prophets, then betrayed and murdered their true messiah. These ideas dominated Christianity's position on Judaism and Jews for 2,000 years. As Jacob Neusner wrote, "At no time before our own century did Christianity contemplate Judaism as an equal, identify in Judaism a medium of salvation distinct from the Church, find in the Torah as read by sages a message both true and also original, or in any way accord to Judaism a place within that tradition of truth that the Church alone nurtured."[2]

This book examines how the Christian Churches initiated and elaborated this theological, mythical, and defamatory image of Jews, Judaism, and Jewishness. Over the past 2,000 years, this anti-Jewish theology—along with its institutionalization and its primitive racism—has influenced, and sometimes determined, the Churches' Jewish policy, that of secular Christian authorities, as well as the behavior of the majorities in the Christian nations in which Jews lived.

Once the break between Christianity and Judaism was made theologically in the first century of the Common Era and finalized politically in the fourth century, the Church attempted to establish its own, unique identity, as independently as possible, from Judaism. To achieve this, the Church cast the Jews in the role of aliens, monsters, pariahs. What most Christian Churchmen taught about the Jews centered around the story of Cain and Abel. But the lesson was not the moral message that we should be our brothers' keepers. Instead, it was the idea, originating in Augustine, that all Jews were Cains—Jewishness and Judaism their stigmata—and the Jews' fate was to wander as suffering examples of what it meant to reject and murder God. For two millennia, Christian theology institutionalized in the Churches has been history's most profound source of antagonism toward Jews, Judaism, and Jewishness; and the Churches have been by far the most significant instruments of Jewish suffering.

It is important to note, however, that throughout Christian history, there existed side by side with *theologia gloriae*, another kind of Christian ideology. It required ethical Christian treatment of all human beings and has been termed the theology of the cross (*theologia crucis*). It is based on Jesus' statement in the Gospel of Matthew (16:24–5): "If any man would come after me, let him deny himself and take up his cross and follow me. For whoever would save his life will lose it, and whoever loses his life for my sake will find it." This belief required the Christian faithful to follow the moral teachings of Jesus concerning all human beings even at the risk of their own lives. Emphasizing the humanity of Jesus, his fears and anxieties as well as his courage and faith, the theology of the cross underscores the solidarity of suffering among all human beings, Gentile and Jew. Analysis of Christians who helped Jews during the Holocaust, for instance, reveals many different motivations for their behavior, but most of these motives derive from the model of human behavior found in the Judeo-Christian morality of Jesus of Nazareth.

During the Holocaust, however, most Christian Churches and most Christians were not adherents of *theologia crucis*, for they stood by in silence, or collaborated, when Jews were taken away by other Christians to be tortured and murdered. This Christian silence was not an ordinary silence. It was special because it justified the murder of Christianity's brothers by the authority of Christian culture (the Churches, their teachings, Christian art and education). Thus many, if not most, Christians became directly or indirectly what they claimed to despise, murderous Cains.

Authorities as dissimilar as Karl Barth and Hannah Arendt believed the connection between the Christian Churches and the Holocaust impossible.[3] The Churches' moral principles, so antithetical to the genocidal morality of Nazi Germany, should preclude, they thought, any connection between Christian precepts and the Final Solution. But those who argue that Christianity's role was indispensable in the etiology of the Holocaust do not refer to Jesus' moral principles, but to a Christian ideology that disdains and hates all things Jewish.

This explains why people—almost all of them born as Christians, baptized and married in a church, and later buried in consecrated soil, coming from a Christian environment, and absorbing a form of Christian culture that condemned Jews—attempted to murder all the Jews of Europe during the Holocaust. And why most other Christians either actively collaborated in this murderous endeavor or tacitly permitted it to happen. In his introduction to *Lessons and Legacies*,[4] Peter Hayes asks two central questions: "How could [Christian] people permit such things?" and "Why did so few brave souls try to intercede?" The answer is that most European Christians—not just in Germany—collaborated actively or passively, to a greater or lesser extent, with what Hitler called for, not so much because of the pressures of fear and anxiety, although these were often present, but because a millennial anti-Jewish Christian ideology had conditioned them into antisemitism.

Nearly every Nazi administrative order—from yellow stars to ghettos, from defamations to deportations, from round-ups to slaughters—had a precedent in the Christian West.[5] Millions of Jews were murdered in Europe before Adolf Hitler was a twinkle in his mother's eye. Jews were condemned as devils from the time of the Church Fathers and regularly massacred from the Middle Ages onward.[6] During the Dreyfus Affair, 20,000 French Catholics—often writing on behalf of their children and their pets—wrote that they planned to flay and butcher and boil the Jewish vampires alive, or bake them in the ovens of Baccarat. A certain Abbé Cros donated three francs for a bedside rug made of "Yids' skins to trample on morning and evenings."[7] They called Jews bugs. A generation later, Germans and their collaborators treated Jews like insects and murdered them in the millions employing an insecticide called Zyklon. The Nazis added a comprehensive organization and a fanatic willingness and technology to follow through to their horrific end the murderous impulses inherent in Christian antisemitism.

Richard Steigmann-Gall points out that Nazism was not essentially an anti-Christian pagan movement,[8] that Christianity played a crucial role in most Nazis' lives and in their Nazism, that Christians believed in the Jewishness of Germany's woes and pointed to a "final solution" of these Jewish-generated problems, that the so-called Nazi pagans—whom many Christian Nazis opposed—were anti-ecclesiastical but not anti-Christian, that Nazi antisemitism fit neatly into Christian antisemitism, that leading Nazis strengthened Protestant Christianity, that in their social policies the Nazis were guided by a Christian ethic, and, finally, that Nazism may have been hostile to the churches but never "uniformly anti-Christian."[9] Many Nazis, both Catholic and Protestant in background, adhered to a "positive Christianity," expressed as early as point 24 of the 1920 Nazi "25 Points"[10] in which they appropriated a divine Jesus Christ as the leading antisemite; they claimed to be authentic Christians above and beyond the artificial division of Catholic and Protestant confessions; "they held that Christianity was a central aspect of their movement [and] shaped its direction, [and] world view."[11] Rosenberg, the Nazi "pagan," called Jesus "a lynchpin of [German] history . . . God of the Europeans."[12]

Ideology was not the only cause of the Nazi Holocaust. A whole raft of political, economic, military, and psychosocial factors also contributed.[13] But the anti-Jewish aspects of Christian thought and theology, the anti-Jewish Christian mindset and attitudes, and the anti-Jewish precedents provided by the churches' historical relationship to Jews significantly conditioned, and may have determined, the plan, establishment, and prosecution of the Holocaust. The churches and their theologians had formulated compelling religious, social, and moral ideas that provided a conceptual framework for the perception of the Jew as less than human, or as inhuman, devilish and satanic, and these churches and theologians had proclaimed Jews traitors, murderers, plague, pollution, filth, and insects long before the National-Socialists called Jews traitors, murderers, plague, pollution, filth, devils, and insects.

Aggravating matters was the racism that crept into Christian anti-Jewish theology. Racism holds that human beings are permanently divided into genetically different groups and that each and every individual within a group manifests the identical intellectual, moral, social, and physical traits of all other members of their group.

Marcel Simon claims that theological anti-Jewishness and racial anti-semitism are totally different, because "from the Church's point of view, at any period, . . . if [a person] was converted, [this person] ceased to be a Jew."[14] But because Christianity has often conceived of the Jews as intrinsically evil and essentially unconvertible, Christian antisemitism has often superseded the Christian sacrament of baptism. From the first centuries of the Common Era onward, many Christians found an inherent theological repulsiveness as well as "a horrible and fascinating physical otherness" in Jews.[15] In 1941, K.E. Robinson, an official of the British Colonial Office, considered the Jews "entirely alien in every sense of the word."[16] Many Church Fathers claimed that every Jew was fundamentally and repugnantly unChristian, if not anti-Christian, and that Jews transmitted indelibly evil characteristics to their offspring. Because of this, the sacrament of Christian baptism could not wash away the "stink of Jewish unbelief." Associating the Jews with heresy, the second-century Christian apologist, Justin Martyr, for example, argued that God had given Moses' Law to the Jews because God wanted to keep the inherently sinful Jews' evil in check. Augustine observed that no Jew could ever lose the stigma of his forebears' denial and murder of Christ.[17] He wrote that the evil of the Jews, "in their parents [*in parentibus*], led to death."[18] His teacher, Jerome, claimed that all Jews were Judas and were innately evil creatures who betrayed the Lord for money.[19] John Chrysostom called Jews deicides with no chance for "atonement, excuse, or defense."[20] Citing Jeremiah 13:23, "Can the Ethiopian change his skin or the leopard his spots?" in the seventh century, Isidore of Seville declared that the Jews' evil character never changed.[21] A Byzantine proverb stated: "when a Jew is baptized, it is as if one had baptized an ass."[22] In the next century, John of Damascus wrote that God gave the Jews the Sabbath because of their "absolute propensity for material things."[23]

These early forms of Christian racism persisted into the Middle Ages. When in 1130 Anacletus II, great-grandson of a converted Jew, was elected pope, Bernard of Clairvaux took the racist position that "it is an insult to Christ that the offspring of a Jew has occupied the chair of Peter."[24] A century later Thomas Aquinas wrote, "The Lord, in order to stir to compassion the Jewish people, naturally inclined to cruelty [*ad crudelitatem pronum*], wished to exercise them in pity even to animals, by forbidding certain practices savouring of cruelty to them."[25]

Fifteenth- and sixteenth-century Spain saw the most extensive development of the Christian racial idea.[26] As Léon Poliakov has pointed out, "Spanish theologians worked out a doctrine according to which the false beliefs of . . . Jews had soiled their furthest descendants . . . . The theologians . . . maintain[ed] that the rejection of Christ had corrupted the *conversos* biologically."[27] Spanish Catholics discovered that once Jews were converted and traditional legal discriminations were removed, many of the perhaps million *conversos* did brilliantly well in Spanish society. Although most *conversos* were of humble origins, others gained influence in the judiciary, the municipal and national government bureaucracy, tax-collection, the army, the universities, and the Church itself. One way for Old Christians to eliminate these successful *conversos* from competition was legally to define their impediment of origin.[28] As Yosef Yerushalmi has noted, "the traditional mistrust of the Jew as outsider now gave way to an even more alarming fear of the *conversos* as insider."[29]

Spanish theologians agreed that "despicable" Jewish ideas and religiously motivated behavior had so corrupted the Jews' descendants that all Jews were impervious to baptism and salvation, so full of the "proverbially 'Jewish' traits of cunning, sharpness, and a boundless lust for money and power defying all moral scruples."[30] By the sixteenth century, when almost all vestiges of Spanish crypto-Judaism were eliminated, King Philip II along with Popes Pius V, Sixtus V, and Clement VIII sanctioned race laws based on the Jews' *mala sangre*, bad blood, whereas Old Christians were exalted by their *limpieza de sangre*, pureness of blood. Spanish race laws—*estatutos de limpieza de sangre*—applied to all *conversos*. A play of the time praised a "noble dog" who smelled out Jews dressed in Christian clothes. In 1604, Father Prudencio de Sandoval wrote, "Who can deny that in the descendants of the Jews there persists and endures the evil inclination of their ancient ingratitude and lack of understanding. [One Jewish ancestor] alone defiles and corrupts him."[31] In 1623, 126 years after the conversion of Portuguese Jewry, a Portuguese scholar held that "a little Jewish blood is enough to destroy the world."[32] Later in the century, after calling the Jews carnal, sensualist, and cruel, a Fr. Francisco de Torrejoncillo warned Catholics: "There is no evil that the Jews do not desire, as they wait for their messiah. . . . To be enemies of Christians, of Christ, and of his Divine Law, it is not necessary to be of a Jewish father and mother. One alone suffices." Christian children must not "be suckled by Jewish vileness because that milk, being of infected persons, can only engender perverse inclinations."[33]

Spanish racism was so widespread that, except for the Jesuits, all the major Catholic orders in Spain in the sixteenth century adopted racist regulations to exclude men of Jewish background from their brotherhood.[34] The Jesuit's founder, Ignatius Loyola, and his immediate successors were traditional religious antisemites who saw the Jews only as theologically, not racially, inferior. Loyola even wished that he had been born with Jewish blood. But in 1608, the Sixth General Congregation of the order voted that no candidate could enter the Society of Jesus unless his Gentile heritage could be traced back five generations. Even the General of the Order could not dispense with this "impediment of origin," which lasted until 1946.[35] Ironically, in 1935 the Jesuits condemned Hitler for the same "purity of blood" principle that the order itself adhered to.[36]

All extensions of Spanish race laws to Catholic orders required papal approval; when the majority of an order wanted to exclude *conversos*, most popes approved the rule.[37] Even the National-Socialist Nuremberg Decrees of 1935 were less exclusive, defining a Jew as having one parent or two grandparents who were Jewish by religious identity. (In the early sixteenth century the warrior pope Julius II reversed the racist momentum, describing such discrimination as "detestable and corrupt and contrary to the wishes of Christ and Paul.")[38] In 1588 and 1600, Popes Sixtus V and Clement VIII approved a Portuguese law that forbade men from Jewish families to be ordained as priests.[39]

These anti-Jewish racial ideas spread from Spain across Europe. Erasmus, the Dutch man of letters and satirist of Church excesses, maintained that Christians must be careful about allowing Jews into the fellowship of the Church, for a fully converted Jew could not exist: he was always pernicious. Erasmus wrote that the Jewish apostate Johannes Pfefferkorn "appears quite typical of his race. His ancestors attacked Christ only, whereas he has [betrayed] Christendom, hypocritically claiming to have become a Christian. . . . This half-Jew has done more harm to Christendom than all the Jews together."[40] One of the authors of *Letters of Obscure Men*, Ulrich von Hutten, described Pfefferkorn in a similar manner. "Germany could not have produced such a monster. His parents are Jews, and he remains such, even if he has plunged his unworthy body into the baptism of Christ."[41] The apostate Jew and Pfefferkorn's friend, Victor von Karben, testified that many Christians mocked Pfefferkorn as a baptized Jew, saying, "anything that is done for you is a waste. [Jews] will never become good Christians. . . . Though you act like a Christian, you are still a Jew at heart."[42] Reuchlin, Germany's foremost student of Jewish literature and mysticism, bucked the trend. His opinion was that "the Jew belongs to God just as much as I do."[43] In his youth a follower of Jerome, Reuchlin at first accused the Jews of hating and persecuting Christians, of unbelief and depravity, and of blasphemy against Christ and Mary. But he later argued that the Jews were citizens of the Holy Roman Empire, they deserved its full privileges and protection, and the medieval accusations of heresy against Jews were false.[44]

Some of Martin Luther's omnibus attack on the Jews was also racist. Luther wrote of the Jews as if they were a "race" that could not truly convert to Christianity. And by making the Jews the devil's people, Luther put them beyond conversion. Trying to convert the Jews, he argued, was like "trying to cast out the devil . . . ."[45] "They have failed to learn any lesson from the terrible distress that has been theirs for over fourteen hundred years in exile. . . . If these blows do not help, it is reasonable to assume that our talking and explaining will help even less. . . . Much less do I propose to convert the Jews, for that is impossible."[46] In a sermon of 25 September 1539, Luther tried to demonstrate through several examples that individual Jews could not convert permanently,[47] and in several passages of *On the Jews and Their Lies*, Luther appeared to reject the possibility that the Jews would or could convert.

> Speaking to them about [the Law, aside from the Ten Commandments] is much the same as preaching the gospel to a sow. . . . From their youth they have imbibed such venomous hatred against the Goyim from their parents and their rabbis, and they still continuously drink it. . . . It has penetrated flesh and blood, marrow and bone, and has become part and parcel of their nature and their life. Dear Christian, be advised and so not doubt that next to the devil, you have no more bitter, venomous, and vehement foe than a real Jew who earnestly seeks to be a Jew. . . . Their lineage and circumcision infect them all. A Jewish heart is as hard as stone and iron and cannot be moved by any means. Even if the Jews were punished in the most gruesome manner so that the streets ran with blood, so that their dead would be counted not in the hundred thousands but in the millions, [it would not be] possible to convert these children of the devil! It is impossible to convert the devil and his own, nor are we commanded to attempt this."[48]

In his trenchant comparison of Spanish and modern German racism, Yosef Yerushalmi observed that these similar forms of racism developed independently and indigenously, oblivious of each other, and that "any hostile conception of the Jews which implies that their negative characteristics are permanent must already be considered as essentially, or at least potentially, 'racial.' "[49]

The Jesuit journal *Civiltà Cattolica*, sponsored and controlled by the Vatican, conducted a racist antisemitic campaign commencing in the last decades of the nineteenth century at least through 1945. In 1880, Father Giuseppe Oreglia wrote: "The Jews—eternal insolent children, obstinate, dirty, thieves, liars . . . barbarian invasion by an enemy race, hostile to Christianity and to society in general. . . . Oh how wrong and deluded are those who think that Judaism is just a religion like Catholicism, paganism, Protestantism, and not in fact a race, a people, a nation. . . . The Jews are not only Jews because of their religion. . . they are Jews also and especially because of their race." Even should they convert, he went on to say, Jews remain Jews for all eternity. The next year, Oreglia added that inspired by the devil, Jews cannot become members of another nation or race, "they are born Jews and must remain Jews. . . . Hatred for Christians they imbibed with their mother's milk."[50] In 1897, Jesuit Father Raffaele Ballerini warned the Catholic world that "the Jew remains always in every place immutably a Jew. His nationality is not in the soil where he is born, nor in the language that he speaks, but in his seed."[51]

When newly appointed Archbishop Theodore Kohn, a Catholic scholar and convert from Judaism, rose to speak at a Catholic Congress in 1896, he was shouted down and the Vatican asked him to resign.[52]

At the time of the Catholic Church's anti-Dreyfus press campaign in 1898, the Catholic daily *La Croix du nord* described Jews as a "race, a foreign race camped among us, a race that has neither our blood nor our ideals, a race that is cosmopolitan by its nature, a race without a country, an intransigent, usurious race lacking moral sense, a race capable of selling and buying anything." Some of these accusations trace back to Jerome's attack on Jews as Judases and later Spanish race laws. Eight years later, another French Catholic newspaper wrote that "the Church of Satan is incarnated in the Jewish race." In 1934, a Polish Catholic journal *Pro Christo* observed that even after seven generations, converted Jews still gave off their "Jew-stink"—a concept related to the association of Jews with the devil, who smelled like feces.[53]

Traditional Christian antisemitism and racism are not incompatible. In the early twentieth century, the Churches denied racist attitudes toward the Jews, assumably because of the Christian sacrament of baptism, but they often acted as if they were judging Jews on the basis of their race. Moreover, many, if not most, racists do not cleanly substitute their pseudo-scientific judaeophobia for their Christian antisemitism. Instead, they build their racist ideology into their already established religious prejudice against Jews. Traditional religious antisemitism has existed side-by-side with racist antisemitism for nearly 2,000 years. For two millennia, through sermons, theological writings, laws, art, and literature, Christian antisemitism has concentrated on the Jews' enduring "sins" and "crimes"—their stiff-necked persistence in their *perfidia*, their greed, their treason, their servitude, their murderous rage at Christ and Christians. On some occasions, Christian racism resulted in mass murder of Jews. The Crusaders and other medieval Christians often massacred Jews, whom they felt were hopelessly unconvertible, without offering them the choice of baptism. These murderers, as had John Chrysostom and Martin Luther, perceived the Jews as irreparably Jewish and worthy of slaughter. The National-Socialists felt the same way and, *mutatis mutandis*, chose the same solution to the "Jewish Problem."

After their study of American opinion in the 1960s, Charles Glock and Rodney Stark discovered that even at a time of growing ecumenical harmony led by the Catholic Vatican II Council, about half of the Americans interviewed—both Catholic and Protestant, both lay and clergy—believed that:

- All Jews were responsible for crucifying Christ, and they could not be forgiven for this act until they converted
- God punishes Jews because they reject Christ
- The Jews are responsible for their own suffering

And the interview respondents were the same people who, associating the Jews with materialism, faulted them for being greedy.

The researchers concluded that far from being exclusively secular, "the heart and soul of antisemitism rested on Christianity." Fully 95 percent of Americans got their secular stereotypes of Jews from the Christian religion.[54] Christianity, as other religions, stands as the focus of prejudice because "it is the pivot of the cultural tradition of a group."[55] This group, Christians, is unlike any other group in Western history; it has been the controlling in-group over the last 1,700 years.

Other studies of prejudice and stereotyping indicate that although the human mind has an inherent tendency to classify, it is not inevitable that people will categorize others by race or ethnicity. This is caused by cultural conditioning. In the England of the seventeenth to nineteenth centuries, for example, Jews were stereotyped as they had been from the time of the Church Fathers, or at least, the Middle Ages and would be into the Nazi era, as: cursed, Antichrists, avaricious, blasphemers, brutes, cheats, circumcisers, cowards, crucifiers, cutthroats, deicides, desecrators of the Host, devils, dogs, fences, parasites, stinking, bleeding, infidels, lascivious infidels, locusts, usurers, murderers, obstinate, stiff-necked, peddlers, perfidious, poisoners, pigs, proselytizers, ravens, reptiles, ritual murderers, serpents, witches, subverters, traitors, thieves, tricksters, cheaters, unclean beasts, wolves. The same pattern occurs in late nineteenth-century France during the Dreyfus Affair.[56] Seventy percent of Americans in the late 1990s demonstrated unconscious stereotyping not because their minds inherently categorized, but because these Americans were emotionally induced to have false memories.[57] It was learned behavior.[58] Many Irish Catholics and Irish Protestants, Hutu and Tutsi, Serbs and Albanians hate and fear each other not because of any inherent predisposition to perceive racial differences, but because of learned religious and political motives.[59]

Christian associations of the word *Jew* with despicable acts and traits have been woven into the languages of the West. In the *Deutsches Wörterbuch*, begun by the Brothers Grimm in 1838 (and completed in 1960), *Jude* was defined as, "Jew: . . . of their evil traits—they are offensive and slovenly, greedy and extortionate[. One says] in a whole variety of idioms—dirty as an old Jew; he stinks like a Jew; . . . to taste like a dead Jew . . . to practice usury, to cheat, to profiteer, to borrow like a Jew . . . ." The dictionary also noted that *Jew* refers to part of a pig's spinal column; *to jew* (*jüdeln*) means to talk, bargain, or smell like a Jew.[60] In the seventeenth century, Littré's *French Dictionary* defined *Juif* as someone who lends at usurious rates or anyone who gains money by means of deceit. In colloquial (Brazilian) Portuguese, as in English, everything associated with Christian is good and valuable, but to *jew* (*judiar*) means to mistreat, to torment, to mock—references to the Gospel interpretation of the Jews' relationship with Jesus. *Judeu*, Jew, means an evil, miserly individual. One saying is that "a Jew, not a Christian, killed my dove, my dove so tame that it would eat out of my hand." *Judiada* refers to inhuman, barbarous, cruel behavior. In popular belief, Jews drank human blood and ate babies. Who would do this? asks the scholar Célia Mentlik, "no other than the Antichrist."[61]

The *Oxford English Dictionary* (*OED*) lists dozens of historical examples of the use of the words Jew and Jewish in English.[62] Fully half the definitions are compound words that are offensive and repulsive. A "jewbush," for example, is one that causes vomiting and death.

The word *antisemitism* has traditionally been distinguished from the term *anti-Judaism* or *Judaeophobia* or *Judenhaß*.[63] *Antisemitism* (*Antisemitismus*) is a nineteenth-century German neologism replacing *Jew-hatred* (*Judenhaß*) in polite discourse and carrying with it overtones of scientific authority, political activity, and racism. Narrowly conceived, *antisemitism* suggests that it was not the religion of the Jews that stirred hostility (anti-Judaism), but biological-race aspects of the Jewish character manifested in their behavior.

But the historical continuity of anti-Jewish ideas and imagery is clear testimony that no essential difference exists between anti-Judaism and antisemitism. One recent author outlined a dozen beliefs of modern antisemites about Jews: (1) conspiracy, (2) intent to conquer the world, (3) desire to harm Christians, (4) immorality, (5) money-grubbing, (6) control of the press, (7) ruination of Christians economically, (8) creation of godless Communism, (9) murder of Christian children and drinking their blood, (10) destruction of the Christian religion, (11) traitors to their nation, (12) Jews must be segregated and their rights curtailed.[64] All these traits—control of the press and creation of Communism could be subsumed under "conspiracy"—are *not* modern but stem from the writings of the Church Fathers and/or the Christian Middle Ages.

Three analogies from the chemical, medical, and biological sciences may clarify antisemitism's ideological functions. First, although they exist within different historical contexts, anti-Jewish ideas, emotions, and behaviors are reactive elements easily combining with other ideologies, such as nationalism, racism, social darwinism, conservatism, fascism, and socialism to form an explosive compound. Second, like a virus, anti-Jewishness rests dormant at different levels of the societal and individual psyche, surfacing especially during the throes of social or personal crisis. Third, although Jews have often been compared to parasites in both medieval and modern antisemitic imagery, antisemitism itself is a parasitic idea, growing more powerful by feeding on the human emotions of fear, anger, anxiety, and guilt.

I use *antisemitism* in the broadest sense, as hostility toward everything Jewish. *Antisemitism* and *anti-Jewishness* are used interchangeably. Both words refer to the irrational dislike or hatred of Jews, the attempt to demoralize or satanize them, the rejection of the validity of the Jewish religion, the Jewish way of life, the Jewish spirit, the Jewish character, and, ultimately, the Jewish right to live. Both antisemitism and anti-Jewishness express themselves as avoidance, antilocution, discrimination, assault, expropriation, expulsion, physical attack, torture, murder, and/or mass murder.[65]

Although an interpretation of Christian theological attitudes toward Jews and its tragic bearing on Jewish history, this book has two underlying moral assumptions. Everyone is responsible for preventing "innocent blood" from being shed and we should love our neighbors as ourselves—both first expressed in the Jewish Scriptures.[66] Shortly after the Holocaust and the Second World War, the German-Christian philosopher Karl Jaspers pointed out that as witnesses to any crime we must intervene, lest we suffer the deepest kind of "metaphysical guilt."[67] Standing in silence before evil is a form of participation in it, although passivity is not as morally tainted as directly committing the evil itself. The Church Fathers who created the theological system in which Jews were cast in the role of the worst people in the history of the world; the medieval theologians and popes and Christian secular authorities who elaborated and enforced this system; American presidents, British government officials, National-Socialists, and, after the fact, those who deny the Holocaust—they all have participated in the evil of antisemitism over the last 2,000 years and have played different, but crucial, roles in the most momentous expression of evil, the Holocaust. We must realize that although history cannot be changed, the past should be studied; what wrong was done in our name should be learned and remembered; we should resolve not to allow such evil ever to repeat itself.

Chapter 2

The Church Fathers

A Sanhedrin of satans, criminals, enemies of God and of all that is decent and beautiful.

—*St. Gregory*

Pagan Attitudes

Although there is some evidence that the word "Jew" was sometimes used as a term of derision before it was redefined by the Latin Church, the pagan Greeks and Romans regarded the Jews little differently from other peoples.[1] There was no intense emotional or ideological hostility. When the Jews of Judea revolted against the Romans, for example, they were treated no more savagely than any other seditious people within the Empire. When Greeks and Romans wrote about the Jews, we find that roughly half of these authors were neutral toward the Jews, one-quarter hostile,[2] and one-quarter friendly. In all, out of 161 Greek and Roman authors who discussed the Jews, only 28 were negative toward them.[3] The pagans respected the Jews' long antiquity, their well-documented history, and their great literature; the Jewish emphasis on family and community; Jewish monotheism, rejection of images, and elevated moral code. The Jews were regarded as a *gens*, a people, an "*insignissima religio, certe licita*," a notable religion, certainly lawful.[4]

Those pagans who were hostile, sniped at the Jews because they were different or annoying. Their antagonism, in contrast to Christian polemic, was social rather than religious, detached rather than emotional, querulous rather than eschatological, literary and aristocratic rather than "propagandistic" (brought down to the level of the common person, and introduced even into the sermons, art, liturgy, and the sacred calendar of the Church), petulant and superficial rather than emotional and dramatic. There was among the pagans no belief or feeling that eternal salvation depended on hating Jews. There was no array of theological ideas supporting, justifying, legitimizing, and sanctifying anti-Jewish hostilities. It is true that there were charges of sacrilege and ritual murder found in a few pagan authors (Manethon, Democritus, Apion), but most of their readers appear to have been incredulous. This was evidenced by the fact that these stories caused no pagan pogroms against the Jews. It is true that we have to await the Middle Ages before the mass murder of Jews results from such charges within Christendom, but it was not till then that Europe was fundamentally Christianized.[5]

Although the pagans did provide some material that Christians directly exploited against the Jews,[6] universal pagan antisemitism is a myth that has served to allow many Christians to exculpate themselves from responsibility for theological antisemitism. Nowhere among the pre-Christian Greeks or Romans do we find the elemental hatred of the Jews that we find first and foremost among Christian writers and Christianized Roman officials beginning with the fourth century, when the conversion of the Empire to Christianity took hold. Christianity brought a completely new factor into the relationship between Roman society and the Jews, and this was *the theological interpretation of Judaism.* Only with Christianity came the idea and sentiment that there was between Christians and Jews a theological war to the death.

Antisemitism in the Christian Scriptures

The profound antisemitism that we know of today began not with the pagans but with the Christian Scriptures. The foundation of the Christian faith was the New Testament, whose writers and later interpreters chose to express an anti-Jewish invective not in their own name but *as if* antagonism to the Jews was part of the mission of Jesus of Nazareth. It was this anti-Jewish "message" that the Church Fathers seized upon, elaborated, and communicated to future generations of Christians. To love Christ came to mean eternal hatred of his alleged murderers and their kin. The Jews had lost their place, their Chosenness to the believers in Christ. The "historical" proof of this theology was evidenced by the fall of Israel: the overthrow of the Jewish king, the fall of Jerusalem, the destruction of the Temple, and the Diaspora of the Jewish people. How could any Christian have ever learned to love the Jewish people, asked Pierre Pierrard, when favorable religious ideas about Jews "were lost in the blood of Calvary"?[7] It is a sad fact that those Christians most immersed in their own Scriptures have often become the most thoroughly bigoted against the Jews.[8]

It is here, first in the Christian Scriptures, and later, in the patristic exegesis of these writings and of the Jewish Scriptures themselves, that Christian anti-Jewish defamation began.[9] For the Christian Scriptures served as great storehouses that the Church Fathers exploited for material to defame the Jewish people. The Scriptural passages damning the Jews as rapacious hypocrites, children of hell and the devil, haters of and rejected by God, and deicides were the focus of nearly all Christian writers of the first seven centuries of the Common Era. Christ-killers was *the* essential Christian accusation against contemporary Jews throughout the patristic period. Tertullian accused the Jews of deicide in twenty passages in ten of his works. Origen, the third-century exegete, regarded the tragic fate of the Jews as due to punishment for their deicide, the culmination of a history of crime, rebellion against God, blindness, hard-heartedness, carnality.[10] In this metaphysical clash of religious identities, "we are now dealing," as Robert Wilken has stated, "with . . . a conflict devoid of reason and logic, a bitter war, the spoils to be nothing less than life itself."[11]

The anti-Jewish portions of the Christian Scriptures that have been seared into the Christian psyche have been these: Acts: "Men of Israel . . . this Jesus, delivered up according to the definite plan and foreknowledge of God, you crucified and killed by the hands of lawless men" (2:22–3); "You [Jews] denied the Holy and Righteous One, . . . and killed the Author of life . . ." (3:13–15); "You stiff-necked people, uncircumcised in heart and ears, you always resist the Holy Spirit. As your fathers did, so do you. . . . And they killed those who announced beforehand the coming of the Righteous One, whom you have now betrayed and murdered . . ." (7:51–2). Matthew: "Kingdom of God will be taken away from you [Jews] and given to a nation producing the fruits of it [the Christians]" (21:43); "Let him be crucified. . . . His blood be on us and on our children" (27:23, 25). Perhaps the most anti-Jewish assertion attributed to Jesus in all Christian Scripture is reported in the Gospel of John: "Your father is the devil and you choose to carry out your father's desires. He was a murderer from the beginning, and is not rooted in the truth . . ."(8:44–5).

Although Paul wrote his epistles before the Gospels were set down, the letters attributed to him seem to encapsulate the New Testament perspective on the Jews. Paul recognized that he was a Jew by birth and background (Phil. 3:5). As to his fellow Jews, Paul sometimes argued that God had "certainly not" rejected them, "God's people" (Rom. 11:1–5). For the Jews possessed "the sonship, the glory, the covenants, the giving of the Torah, the worship, and the promises. The fathers are theirs, and of them is the Christ, as a human person" (Rom. 9:1–5). John Gager, Clark Williamson, and Norman Beck, among others, have argued that despite his ambivalent feelings toward Judaism, Paul saw that the "old" Law was valid for the Jews, but the "new Law" of Christ was legitimate for Christians. Jews and Christians are regarded each as a People Chosen by God in its own way (Rom. 11:16). Gager concludes that Paul's Gospel "did not entail repudiation of the legitimacy of Israel or the Torah."[12]

But many of the writings attributed to Paul discounted much that was essential in Judaism and introjected a high level of emotional polemic into the controversy between Jews and Christians. The words attributed to him often expressed a hatred for Judaism in general and for the very Torah that made Jews Jews. As David Flusser has recently written, "Opposition among Christians against 'legalism' demonizes the Jewish law and makes the Jewish religious way of life monstrously hostile to God."[13] (1) In Romans ll:25, Paul posited that Judaism was *not* as "valid" as Christianity, for in the last days the Jews would reject the religion of their fathers and allow themselves to be converted to Christianity. (2) Romans argued that in Judaism faith and Law are separate, that is, it is a "legalistic" religion. Paul took the supersessionist position that "What once had splendor has come to have no splendor at all, because of the splendor that surpasses it" (3:4–16). (3) Of the Torah, which Jews considered indispensable to their Jewishness, Corinthians stated that "the written code kills," it is "the dispensation of death," "whenever Moses is read a veil lies over their minds" (2 Cor. 3:4–15). (4) It was implied that the Torah was the essential basis for Jewish conduct (Rom. 2), yet its physical sign, circumcision, was a "mutilation" (Phil. 3:2) and the whole Torah "a curse" (Gal. 3:13). (5) In Philippians 3:8, Judaism was referred to as "refuse" or "dung." (6) In the same letter (3:2–8), the Jews, or Judaizing Christians, that is, Christians who were backsliding into Jewish customs, rituals, and the like, were called "dogs" and "evil-workers." (7) Paul argued that "as regards the gospel [the Jews] are enemies of God" (Rom. 11:28–9). (8) In 1 Thessalonians (2:13–16), the whole Jewish people were defamed as murderers of Christ. (9) 2 Thessalonians 2:3 noted that the great theological enemies of Christ were not the Gentile Greeks or Romans, but the Jews. The Antichrist was the false Jewish Messiah who worked through Satan. He was sent to the Jews by God specifically because they have refused to believe in the true Messiah, Jesus Christ.[14]

In some of these passages Paul may in fact be polemicizing only against Judaizing Christians, but in the exegesis of the Patristic theologians and of later Christians no such distinction was made. The key to understanding the fourth-century Cyril of Alexandria, for example, is to realize that he adopted Paul's interpretations of Moses and the Law as death; Christ as life. Cyril wrote that Paul "considered [the Law] rubbish," and gained Christ. Paul's testamental antagonism to Jews and Judaism, as in the Gospels and Acts, as well as the references in Revelation to the Jewish Jews as a "synagogue of Satan" (2:9, 3:9), have provided a justification for the antisemitism of most Church Fathers and later Christian theologians.[15] These are the basic myths that have dominated the Christian West's perception of the Jews for two millennia. Even when we factor in the competition between Christianity and a thriving Jewish community that denied the validity of the new religion, it is still obvious that the antisemitic interpretation of the New Testament by the Church Fathers is the main root of antisemitism. ("Church Fathers" is an inexact term referring generally to Christian thinkers, theologians, writers, teachers, and bishops from the third to the eighth centuries. Before this period, the Apostolic Fathers flourished until the early second century; the Christian Apologists from then until the early third century.)[16]

Christian Anti-Jewish Writings

Especially before the fourth century, Jews presented a "political" threat and served as a theological rival to Christianity. A thriving Jewish community contested the tenets of the new Christian religion, and the Jewish religion was legally sanctioned within the Roman Empire. Aspects of Judaism attracted both pagans and Christians. Judaizing—the movement of Christians or pagans toward Judaism—greatly concerned Church authorities. In the Christian centers of North Africa and the Near East, numerous pagan "God-fearers" as well as many Christians observed the Jewish Sabbath, listened to the rabbis' preaching, celebrated Jewish festivals and customs, and practiced circumcision. Jews welcomed these judaizers as "strangers within the gate." A Jewish state, that of the Himyarites, existed on the southern Arabian Peninsula and may have controlled the eastern shore of the Gulf of Suez, threatening the Christianized Empire itself between the fourth and sixth centuries.[17]

Christian theologians before the fourth century expressed a much greater ambivalence toward the Jews than the later Church Fathers, who were more hostile.[18] Having to fight off Judaism's legal, political, and theological challenge, defeat the pagans, and establish Christianity's credentials required the early fathers to moderate their opposition to the Jews. Because historically minded pagan thinkers ridiculed Christianity's rejection of the age-old Jewish Torah as interpreted by the Jews,[19] many early Church Fathers moderated their antagonisms and emphasized the connections between Judaism and Christianity when they discussed Judaism with the pagans who revered Judaism's longevity and opposed the Christian mission.

Christian theologians pillaged Judaism to supply Christianity with an unimpeachable history and a prestige the new Church would not have otherwise possessed. The fathers argued that Christianity fulfilled the prophecies in the Jewish holy books, especially those foretelling a Messiah. The fathers condemned the Jews' blindness in refusing to accept Christianity as the one true faith but recognized that the Jews "had received the call of God." In the third century, Origen praised the Torah as inspired by the holy spirit of God and extolled Jewish moral-legal principles upon which most Christian thinking was based in his tract against the pagans, *Contra Celsum*. But his praise of the Jewish foundations of Christianity was restricted to the Jewish past looked forward to the possibility of Jewish conversion.

But when Christian authors wrote about Jews, Judaism, and Jewishness for a Christian audience and not in argument against pagans, nearly all of them demonstrated an almost compulsive hostility to the Jewish spirit. Western and Eastern theologians expressed a hatred and fear of the Jews.[20] These writers throttled whatever neighborly relations had existed between Christians and Jews, clearly crossing the boundary from reasoned debate to emotional polemic.[21]

Christian writings were part of a theological war to the death, and beyond. Although several Church Fathers knew individual Jews, they portrayed Jews as satanic adversaries.[22] They imagined that Jews insulted Christ and the Blessed Virgin each day in their synagogue prayers. For these crimes, Christian theologians argued, Jews must suffer continual punishment on earth and eternal damnation in the afterlife, unless they sought salvation through the one true faith, Christianity. They proclaimed that the Jews are, have always been, and will always be, paragons of evil.

Christian theologians depicted the Jews as hateful to make Judaism repulsive to the Christian faithful or pagans. The stubborn persistence of Judaism and Jews constantly questioned Christian claims of earthly and spiritual triumph. The intensity of anti-Jewish language in portions of the Christian Scriptures and from almost every Christian theologian from the Church Fathers forward—their writings became almost as authoritative as Scripture—was both the cause and the result of this concern for the potential loss of Christian and pagan souls. In the writings and sermons of these early Christian propagandists, no evil was too great for the Jews not to revel in, no crime too appalling for the Jews not to rejoice in. Hilary, bishop of Poitiers, wrote in the fourth century, Judaism was "ever . . . mighty in wickedness; . . . when it cursed Moses; when it hated God; when it vowed its sons to demons; when it killed the prophets, and finally when it betrayed to the Praetor and crucified our God Himself and Lord . . . . And so glorying through all its existence in iniquity . . . ."[23]

The Transmogrification of Jewish Values

The paramount tasks of the emerging Christianity were to overthrow the theological dominance of Judaism, to establish itself as a separate and self-sustaining religion, and to situate Christians as the new and authentic chosen people.[24] The Christian Fathers believed that they were fighting a war with the Jews in which only the victor would reap the reward of eternal life; Judaism was a standing insult and threat to Christianity's image of itself.[25] Jerome once wrote to Augustine that if converted Jews were allowed to practice even one fragment of their former religion, "they will not become Christians, but they will make us Jews. . . . The ceremonies of the Jews are pernicious and deadly; and whoever observes them, whether Jew or Gentile, has fallen into the pit of the devil. For Christ is the end of the Torah . . . ."[26] John Chrysostom was blunt about it, "Don't you realize, if the Jewish rites are holy and venerable, our way of life must be false. . . . The Jews . . . pay honor to the avenging demons, the foes of our life."[27]

Through anti-Jewish theological myths and defamations, the Church Fathers pictured the Jews no longer as the chosen people, no longer heroes of holiness and moral living; they were instead the earthly representatives of the powers of evil. The Church countered the Jewish belief in the possibility of Gentile salvation with assertions that only Christian nations should exist and that salvation was inconceivable outside the Church.

To provide the Church with a clear identity for itself, Christian theologians transvaluated Judaism. The Church Fathers turned Jewish values and practices on their heads by misrepresenting them as their opposites. Cyril of Alexandria wrote, "Note that the shadow of the law is reversed and the ancient things of the law are made ineffective."[28] Christian theologians attacked the essential values and practices of Jewishness—the Jews' Covenant with the One God, their Chosenness, circumcision, ethical laws, God-wrestling, Messiah, dietary obligations, Sabbath, holy days, patriarchs, and Holy Scriptures. The Church Fathers did this by reinterpreting, modifying, and adapting them to fit the requirements of the Christian self-image and theology. In effect, the Latin and Greek Fathers said to the Jews, "We'll take your God, your Scriptures, your Messiah, and some of your Law; as for you, you are disinherited, cast into a limbo, and your survival serves only as a warning of the consequences of obdurate wickedness."[29]

The Jews claimed to have an intimate relationship with a spiritual God, who they believed had created all humanity, and with whom they had made a Covenant. In it, they agreed to fulfill moral and ritual obligations; in return, God would make them the Chosen People, the males marked by ritual circumcision. (In Hebrew, being holy involves being set apart from others. Moreover, every people in one way or another sees itself as chosen.)

It was believed that the souls of all Jews were present at Sinai and that for all eternity all Jews were committed to this contract with God. The Covenant obliged the Jews to act as a kingdom of priests and a holy nation, setting an example by the conduct of their lives and helping human beings move toward righteousness.[30] Jews saw themselves as living witnesses to God's moral purpose for mankind, even if, as God's servants, they had to suffer and die for it.

Christians agreed that all Jews were eternally responsible for Jewish behaviors. But Christians believed that the Jews were not God's allies but God-murderers—people who first rejected and then slew God, who had come to earth in the form of Jesus Christ. Through this act, the Jews incurred the wrath of God the Father and, according to Augustine's myth of the Witness People, had to live throughout history suffering for their sacrilegious crime. The Epistle of Barnabas, written around the year 100, noted that the Jews had forfeited their Covenant with God just after Moses had received it, and Christians had replaced them as God's Chosen. Because the Jews had immediately turned aside from the Law, Moses threw down the tablets of the Law. "That Covenant of theirs was smashed to pieces, so that the seal of the Covenant of Jesus the Beloved might be stamped on our own hearts . . . ."[31]

Christian writers claimed that circumcision no longer symbolized the Jews' Covenant with God; instead, it marked the Jews as devils or Cains. In his letters, Paul asserted that the Jewish circumcision of the body could not compare with the Christian circumcision of the spirit.[32]"It is not the children of the flesh who are the children of God, but the children of the promise are reckoned as descendants. . . . 'The elder will serve the younger.' As it is written, 'Jacob I loved, but Esau I hated.' "[33] Justin Martyr wrote in his *Dialogue with Trypho*: "Circumcision was given to you as a sign that you may be separated . . . from us, and that you alone may suffer that which you now justly suffer, and that your land may be desolate, and your cities burnt with fire. These things have happened to you in fairness and justice." The fourth-century theologian Ephraem of Syria called the Jews circumcised dogs; John Chrysostom called them circumcised beasts.[34] Aphraates held that circumcision was not necessary to mark out the authentic Israel, since Abraham had not been circumcised. Tertullian suggested that God intended that the circumcision would identify the Jews so that they could never reenter Jerusalem.[35]

Ethical laws abound in Jewish Scripture and the Talmud, the two foundations of Jewish life. One has only to consult any Jewish prayerbook to get a feel for the dominance of ethical principles. These include the abhorrence of shedding innocent blood (the consumption of any blood was seen as an abomination, since "blood is life")[36] and the condemnation of those who refuse to be their brother's keeper.[37]

The Christian Fathers ignored the Jews' ethical principles and focused instead on the "blood-guilt" of the Jews in connection with the so-called murder of their own prophets, the crucifixion of Jesus, and their alleged call for Jesus' blood to be on their heads and the heads of their children. The Church Fathers based much of their attack on Matthew 23 (although only "scribes and Pharisees" are mentioned here, all Jews are implicated), which condemned the Jews as evildoers from the beginning of their history. "You are descendants of those who murdered the prophets. . . . How can you escape being sentenced to hell? . . . [U]pon you may come all the righteous blood shed on earth, from the blood of righteous Abel to the blood of Zechariah . . . whom you murdered between the sanctuary and the altar. Truly I tell you, all this will come upon this generation."[38] Far from treasuring life, the Jews were accused of committing incest with, and murdering, their own children.

Another Jewish ethical concern was God-wrestling, a tradition in which any Jew could remind God of God's moral law.[39] After Jacob wrestled with God in the form of a man, God first called the Jews *Israel*, which literally means "the one who strives [wrestles] with God."[40] The Church Fathers transformed God-wrestling into the myth of Jewish revolution. Jews revolted against traditional moral order and legitimate political authority and were parasites in every society in which they lived. The Christian Fathers accounted Judas Iscariot, the archetypal Jew and the least redeemable human being in sacred history, as a materialistic betrayer of God. The Fathers treated the Jews as enemies of human kind rather than valuers of human life.

The Jews also maintained that they were awaiting the true Messiah, God's agent whose arrival would be disclosed by the establishment of justice, peace, and harmony for all living beings. Until he came, Jews themselves would work to change the world for the better. However, Christian defamers replied that the predicted Messiah had already come; he was the redeeming God, Jesus Christ. That the Jews still awaited their Messiah, an essential aspect of their self-identity, became further reason for Christians to oppress them. As Jacob of Serug put it in the fifth century, the prophecies would not be twice fulfilled; Christ "gave no opportunity for another to come."[41] Because they continued to deny Christ while awaiting their false Messiah, the Jews were considered allies of the Antichrist, as well as deicides.[42]

The Torah, Talmud, and Jewish Apocrypha agreed that the pig was the symbol of filth and of all that was abominable. Since they associated the sacrifice of pigs with idolatrous worship, Jewish authorities condemned even swine farmers. For a Jew, the refusal to eat swine, even when threatened with death, became a test of one's Jewishness.[43] Yet Christian theologians charged that Jews not only associated with the pig, but worshipped it.

Observance of the Sabbath had many meanings for Jews: a way to sanctify time, to reestablish community, and to contemplate one's life and one's relationship with God. But in the eyes of early Christian writers, the Sabbath became a time when the Jews plotted their evil deeds. John of Damascus (675–749) held that God gave the Jews their Sabbath because of their "grossness and sensuality . . . and absolute propensity for material things."[44] Cyril of Alexandria, a "fanatically strict patriarch" who himself led riots against the Jews and expelled them from his city,[45] stated that the Jewish "Sabbath was rejected by God."[46] Early Christians secularized the Jewish Sabbath and replaced it by a new Sabbath, the Lord's day, Sunday.

Believing that once Christ had come Jewishness should have ceased to exist, Christian theologians also ridiculed other holy days, by which Jews mark the sacred calendar and remember their historical and religious identity. The Jewish observance of their holy days was considered an insult to Christianity. Yom Kippur was a "wicked and unclean fast." The devil himself summoned Christians to witness the Jewish Feast of the Trumpets (Rosh Hashanah). Jewish festivals were "a table of demons because they slew God."[47]

Christian theologians also systematically reinterpreted the meaning of the synagogue—treated by Jews as a house of fellowship, learning, and prayer. Perhaps because it symbolized Judaism in general as well as representing the various Jewish communities in particular, several Church Fathers attacked the synagogue in the most lewd and vicious terms imaginable. John Chrysostom, Jerome, and Ambrose called the synagogue a "whorehouse,"[48] a "den of vice,"[49] "a refuge of insanity."[50]Christianized Roman law reflected these attitudes as well.

By co-opting the Jewish patriarchs, saints, and true believers in God all the way back to Adam, the Church dispossessed the Jews of both their past and their future as the Chosen People. The Church Fathers considered the admirable Old Testament prophets such as Abraham and Moses in reality Christians; only those Scriptural figures who had sinned were considered Jews who often expressed their evil by murdering the prophets: Paul, Eusebius, Augustine, Justin Martyr (who may have been the first to accuse the Jews of deicide);[51] and Aphraates substituted the Christians for the Jews as the authentic "descendants of Abraham."[52] Justin asserted that the true Israel of Judah, Jacob, Isaac, and Abraham were those "whom the crucified Christ led to God."[53] Aphraates misquotes the Gospel of John, claiming that Jesus said to the Jews, "You are the children of Cain, and not the children of Abraham." (In actuality, John's Gospel states: "If you were Abraham's children, you would be doing what Abraham did, but now you are trying to kill me . . . . You are from your father the devil, and you choose to do your father's desires. He was a murderer from the beginning and does not stand in the truth, because there is no truth in him.")[54] Augustine made the argument that the important Jewish prophets were not really Jews—they were the earliest Christians—and that Christians, not Jews, were the true people of salvation, the authentic Israel.[55] Pseudo-Cyprian stated: "Moses they [the Jews] cursed because he proclaimed Christ, . . . David they hated because he sang of Christ, . . . Isaiah they sawed asunder shouting His glories, . . . John they slew revealing Christ, . . . Judas they loved betraying Him."[56]

The Church Fathers argued that only Christian history—having begun at the beginning of time as recorded in the Jewish Scriptures—made sense of Jewish history. "The one and true God, creator of goodness . . . is the author of both Testaments; but what is New is predicted in the Old Testament, and what is Old is revealed in the New."[57] Pseudo-Barnabas warned that Christians must not regard the Jewish Scriptures as belonging to both Jews and Christians. "It is ours only; for they have lost forever that which Moses received." Justin Martyr speaks to a Jew of "your scriptures, or rather, not yours, but ours."[58] Augustine asserted that Jewish moral and historical blindness prevented them from recognizing Jesus as the Messiah and so God condemned them to act as slaves who merely "carry their masters' books. The slaves gain nothing by their carrying; but the masters profit by their reading."[59] Had the Jews realized to what use their Law and the writings of their prophets would be put by Christians, "they would never have hesitated," said Irenaeus, the second-century bishop of Lyons, "themselves to burn their own Scriptures."[60]

Tertullian: Christian Self-Identity and the Jews

In their ideological assault on the Jews, early Christian theologians interpreted the Roman victory over the Jews, the loss of the secular Jewish kingdom, the Jews' holy capital Jerusalem, and the land of Israel as evidence of God's abandonment of the Jews and of their eternal punishment in this life and the next. God was always pictured as "in there punching" on the side of Christianity and Christians against Judaism and Jews.[61]

Tertullian (d. 225) was the outstanding North African theologian of the second and third centuries, called "the Father of Latin theology."[62] With anti-Jewish diatribes contained in 27 of his 32 extant works, it is safe to say that theological antisemitism was a precondition of his religious beliefs, Tertullian regarded the ramifications of the Jewish challenge as a great danger to the emerging Christian identity. He argued that Jews were "the very anti-type of true virtue," since they preferred their prophets to Christ, insulted and persecuted Christians, and rebelled against the good.[63]

Tertullian constructed an essentially anti-Jewish reality. His writings are replete with attacks on the Jewish people for a whole panoply of "crimes"; 23 categories: from "bad habits" such as clinging to the past and prophet bashing to deicide, worst of the Jewish sins. In *Adversus Judaeos*, Tertullian noted: "Thus the whole synagogue of the sons of Israel slew him, saying to Pilate, who wanted to release him, 'His blood be on us and on our children.' "[64] In *De Oratione*, he wrote that "though Israel may wash all its members every day, it is never clean. Its hands . . . are always stained, covered forever with the blood of the prophets and of our Lord himself."[65]

Like most of the fathers, Tertullian's anti-Jewish conclusions were often both emotional and cruel. In his *De Spectaculis*, he gloated and exulted, imagining how Jesus would punish the Jews for having "thrown God, i.e., Christ, out." Israel was not merely *extra ecclesiam* (outside the Church); it was "*extra Deum*" (outside of God). "I . . . would prefer to turn an insatiable gaze on those who vented their rage on the Lord. 'This is he,' I will say, 'the son of the carpenter and the harlot . . . . This is he whom you [Jews] purchased from Judas, this is he who was struck with the reed and fist, defiled with spittle.' "[66]

Jerome: The Jews as Judases

The most influential Church Fathers of the fourth and fifth centuries attack the Jews in the same theological terms as Tertullian. Sophronius Eusebius Hieronymus, called Jerome (d. 420), was a European priest who spent most of his time in Antioch and Bethlehem. His translations served as the basis of the Vulgate, the Latin Church's official Bible. Calling the Jews "poison snakes" and "wolves," he argued that God had given the Jews their Law deliberately to deceive them and lead them to their destruction.[67] He envisioned a Jewish Antichrist who would become king through fraud, conquer the Romans, and persecute Christians.[68]

Jerome's most important contribution to theological antisemitism was his identification of the Jews with Judas and with the immoral use of money: two themes that would bedevil Christian–Jewish relations for a millennium. From the beginning of the Christian era, Jews had been associated with materialism and "carnality." In emphasizing this corporeal nature of the Jews, Paul noted the difference between the physically circumcised Jews and the real Israel, the spiritually circumcised Christians.[69] Jews are attached to the Old Law, which means "the flesh" and "sinful passions," but Christians are dead to this "old written code" and slaves only to "the new life of the Spirit."[70] Jews walk "according to the flesh," Christians walk "according to the Spirit." (In Gal. 5:19–23, Paul indicates what living in the flesh means: "fornication, impurity, licentiousness, idolatry, sorcery, enmities, strife, jealousy, anger, quarrels, dissension, factions, envy, drunkenness, carousing." Whereas the spiritual life consists of "love, joy, peace, patience, kindness, generosity, faithfulness, gentleness, and self control.") The Jewish mind is set on death, but the Christian mind is set on the Spirit, which "is life and peace." The Jews are "hostile to God," whereas the Christians "have the Spirit of Christ."[71] Paul argues that "what is truly human [is] the human spirit."[72]

The first-century Epistle of Barnabas noted that the Jews had "carnal instincts."[73] Matthew, Mark, and Luke indicate that the chief priests promised Judas money to betray Jesus.[74] Although only Matthew specifies the amount, 30 pieces of silver, the Gospels clearly identified Judas with the worst kind of materialism, taking money to betray one's savior. Aphraates derided Jewish "sensuality and carnal desires" and cupidity.[75]

All that remained was for all Jews to be identified with carnality and Judas. Jerome associated materialism with Judas' sin, which, with his punishment, symbolized Jewish behavior and fate.

> "Christ is saying: 'Judas betrayed Me, the Jews persecuted and crucified Me' . . . . In particular, this is the story of Judas; in general it is that of the Jews[76] . . . . Judas, in particular, was torn asunder by demons—and the [Jewish] people as well. . . . Judas is cursed, that in Judas the Jews may be accursed." "[Even] the repentance of Judas became worse than his sins. [Just as] you see the Jew praying; . . . nevertheless, their prayer turns into sin. . . . Whom do you suppose are the sons of Judas? The Jews. The Jews take their name, not from Juda who was a holy man, but from the betrayer. . . . From this Iscariot, they are called Judaeans. . . . Iscariot means *money and price*. . . . Synagogue was divorced by the Savior and became the wife of Judas, the betrayer."[77]

In another emotional passage that may have served as the basis of a portion of the later Good Friday liturgy called the Reproaches, Jerome contrasted the gifts God had given the Jews with the "evil" with which the Jews repaid them.[78] This homiletic assault went on for 4,000 words in this vein: "My enemies are the Jews; they have conspired in hatred against Me, crucified Me, heaped evils of all kinds upon Me, blasphemed Me." Jerome ends his sermon by asking his parishioners to forgive the Jews and to pray for their souls, but his central theme is the complicity of the Jews in the murder of Jesus, their true Messiah.[79]

Jerome's ambivalence, some could call it hypocrisy, about Jews, damning them and then asking Christians to forgive them, is a long-lived characteristic of the Church's position on the Jews. We will see that the medieval papacy protected Jews but not consistently, generally following Augustine's advice that Jews were to suffer but not be killed. In 1898, the Vatican's newspaper *Osservatore Romano* attacked Jews as vampires thirsting for Christian blood and conspiring to destroy Christianity, suggests that "a healthy antisemitism . . . is nonviolent."[80] Later, during the Holocaust, the Vatican and other Roman Catholic representatives in Europe supported legal discrimination against Jews, but always done "with justice and charity" and although occasionally condemning antisemitism at the same time justified it as "self-defense."[81]

Augustine, Ambrose, and others grudgingly conceded the chastity, charity, and generosity of the Jews.[82] But this praise is very rare. As Eliezer Berkovitz has noted, "charity asked for a people that in the same breath is called fallen and faithless has little effect in history."[83] The words of Jerome that ring down through history are not his pleas for forgiveness, but his complaint to Augustine, whom he believed was too lenient with the Jews, that "the ceremonies of the Jews are harmful and deadly to Christians, and . . . whoever keeps them, whether Jew or Gentile, is doomed to the abyss of the Devil."[84]

The impossible mixture of "charity" and discrimination would afflict the Church, its popes and councils for centuries. As kind a prelate as Cardinal Mattei of Ferrara in 1781, under Pope Pius VI, prescribed that Christians in his diocese could not hire themselves out to Jews, consort with them, be medically treated by them; that Jews must be marked with stigmatic signs and stay indoors during Holy Week—all this to be achieved "with Christian charity."[85]

Augustine: The Jews as Witness People

Denying the vulnerability of the Jews—most negative Christian perceptions of Jews have existed independently of what Jews themselves have actually done and of a Jewish presence at all[86]—Christian theologians transformed the image of the Jewish people into a symbol of evil that theologically justified the elimination of Jews through segregation, expulsion, conversion, or sometimes murder, the Jews' loyalty to their own beliefs being considered an insult and a danger to Christianity's image of itself.[87] The Christian dilemma was that because Christianity had no history without Judaism, it had to be preserved, but in a state where it could do no "harm" to Christianity, like a body in suspended animation. Instead of destroying the Jews, thereby cutting of Christianity from its roots, the Church Fathers asserted that the Jews were "an indispensable reference group, enabling Christians to know themselves as Christians and to incarnate good by contrast with evil."[88]

Aurelius Augustinus—Augustine (d. 430)—bishop of Hippo Regius and a great doctor of the Church, was a prolific, inventive theologian. Many of his ideas on the Jews reflected the ambivalence expressed in Paul, who regarded Jews as both the thesis and antithesis of Christianity. Augustine's Christian love for Jews was intended to result in their conversion, nothing else: "Let us preach to the Jews, whenever we can with a spirit of love. . . . We shall be able to say to them without exulting over them . . . 'Come, let us walk in the light of the Lord.' "[89]

He was infuriated by three issues: (1) the Jews' resistance to conversion, (2) the fact that they were the first keepers of Scriptures sacred to Christians, and (3) the attraction they held for the Christian faithful.[90] To counter his objections, Augustine developed these doctrines: (1) "there is no salvation outside the Church," which asserted that Jews who refused to convert were damned to hell; (2) proof of the Jews' inherent evil existed in their own sacred writings and the Jews who rejected Jesus as the Christ were not true Jews, Christians were the authentic Jews;[91] Christians who observed even the smallest Jewish rituals, that is, judaized, were heretics.[92]

A disciple of Ambrose, bishop of Milan, Augustine may have inherited his teacher's virulent anti-Jewishness. Ambrose had identified the Jews with the fratricide Cain, who was condemned to walk the earth suffering because of his fratricidal sin. (In later tradition, Cain became associated with the Devil.)[93]

> These two brothers, Cain and Abel, have furnished us with the prototype of the Synagogue and the Church. In Cain we perceive the parricidal people of the Jews, who were stained with the blood of their Lord, their Creator, and . . . their Brother, also. By Abel we understand the Christian who cleaves to God.[94]

Augustine also held that, as Cains, the Jews should wander endlessly while suffering punishment for their murder of Jesus. Just as Cain had served as a warning to those tempted by murder, so the Jews should wander as suffering examples to Christians who were tempted to revolt against Christ. After comparing the Jews to poison snakes who are deaf to Christianity's "enchanting" message, Augustine interprets Psalm 59 against the Jews.[95] In this psalm, David asks God's protection from his "bloodthirsty" enemies "who treacherously plot evil." To confirm the Lord's triumph (and his own), David implores God to "consume" these enemies in "the sin of their mouths, the words of their lips, . . . their pride[,] cursing and lies," not to kill them but to "bring them down." Augustine believes that this psalm proves that

> The Jews have been scattered throughout all nations as witnesses to their own sin and to our truth. They themselves hold the writings that have prophesied Christ.

> If a pagan doubts Christ, we can prove his Messiahship because he was predicted in the writings of the Jews themselves a long time ago. And so by means of one enemy [the Jews] we confound another enemy [the pagans]. "Scatter them abroad, take away their strength. And bring them down O Lord."[96]

In his reply to Faustus the Manichaean, Augustine continues his analogy between Cain and the Jews, arguing that the Jews were materialistic and evil deicides who will be punished forever, until they see the light and convert to Christianity. "Not by bodily death," Augustine wrote,

> shall the ungodly race of carnal Jews perish. . . . To the end of the seven days of time, the continued preservation of the Jews will be a proof to believing Christians of the subjection merited by those who, in the pride of their kingdom, put the Lord to death. . . . "And the Lord God set a mark upon Cain, lest any one finding him should slay him." . . . Only when a Jew comes over to Christ, he is no longer Cain.[97]

By identifying the Jews with Cain, Augustine turned the Jewish historical and moral mission on its head. The Jews were no longer the divinely chosen witnesses to God's moral message, they instead were now a sinful "Witness People" who would prove to the pagans the melancholy fate that awaited those who opposed Christ—a concept that legitimized and sanctified the suffering enslavement of Jews to Christians.[98] Although Augustine's doctrine established a limit to Jewish suffering (when Donatus, Proconsul for Africa in 408, interpreted a new law [*Codex Theodosianus* 16:5:44] penalizing heretics and Jews who did "anything contrary and adverse to the Catholic sect" with death, Augustine convinced him to mete out a lesser punishment),[99] his central lesson was not that Christians should act as their Jewish brothers' keepers, but that Christians should act with hostility toward the Jewish Cains—the same way Cain himself had behaved toward his brother.

Augustine's metaphor that Jews were branded by an indelible mark of Cain would continue to haunt Jews living amid Christian peoples. Some Christian theologians even saw Jewish circumcision as Cain's mark. At one point, Augustine seems to indicate that the mark of Cain was the Torah itself. Augustine follows a reference to Cain with the observation that "[T]he Jewish nation . . . has never lost the sign of the law by which they are distinguished from all other nations and peoples. No emperor or monarch who finds under his government the people with this mark kills them . . . ."[100] His idea would justify the medieval papacy's policy of Jewish degradation. Pope Innocent III and the Fourth Lateran Council insisted that secular governments order the mark of Cain sewn onto Jewish clothing. The mark was later transmuted into a burning cross on the forehead of the "Wandering Jew."[101]

Ambrose: Burn Down Their Synagogues!

Once Christianity came to dominate the Empire in the fourth century, attacks against synagogues dramatically increased. The position of even the early Christian emperors had been that the Jews should be allowed to use their synagogues for constrained worship, for these privileges had traditionally been theirs. But we know of attacks on synagogues often with the instigation of the local bishop or as the result of anti-Jewish sermons, and resulting in the forced conversion, exile, or death of the Jewish congregants at Rome between 383 and 388,[102] Callinicum in 388, Edessa in 411–12, Alexandria in 414, Magona on the island of Minorca in 418,[103] and in Syria, Palestine, and Trans-Jordan in 419–22.[104] The Imperial edicts "protecting" synagogues were often based on the "supplications of Jewish delegations," limited by concessions to the Church, and characterized by the absence of penalties for their violation. They had to be repeated by Emperors Valentinian I, Valens, Theodosius I, Theodosius II, Arcadius, and Honorius in 368–73, 393, 397, 412, 420, 423 (*C.T.* 7:8:2, 16:8:9, 16:8:12, 16:8:20, 16:8:25; *C.J.* 1:9:4). The law of 412 interestingly reflects the imperial petulance of Honorius and Theodosius II over having to repeat such laws: "it would seem that enough had been legislated on this matter in general constitutions by past Emperors" (*C.T.* 16:8:20).

An important reason for the assaults on the synagogue was that to Christian theologians, the synagogue represented Judaism, the hated rival of Christianity. Let us recall Jerome's opinion on the synagogue: "If you call it a brothel, a den of vice, the Devil's refuge, Satan's fortress, a place to deprave the soul, an abyss of every conceivable disaster or whatever else you will, you are still saying less than it deserves."[105] To him, the Jewish hymns sung in the synagogue were the "braying of jackasses."[106] To destroy or damage the synagogue building, to transform it into a church, or to prevent the construction of a new synagogue were means for the politically empowered and theologically motivated Church to persecute the Jews, lower their prestige, and destroy their attraction for the Christian faithful. Gathering inspiration from the Church Fathers, Christian prelates indoctrinated the faithful so that their policies in regard to the Jews would be effectuated. Ironically, by the fourth century, the Church had less to fear from the Jews than ever before. Jewish attacks against Churches occurred rarely, and then apparently only in reaction to Christian destruction of synagogues.[107] Without equal legal status, Jewish responses to Christian depredations were manifested mostly through ridicule, for example, in the caricature of the life of Jesus, *Toledot Jeshu*, elements of which existed from about the fifth century, although it was not finalized until the tenth. Jews also exhibited their frustration by making anti-Christian puns, calling the Gospel "a revelation of iniquity," and the apostles "apostates."[108]

Several synagogues were transformed into churches, the consecration of the building often taking on the semblance of an attack on Judaism and proclamation of the victory of Christianity. A special Mass using the Gelasian Sacrament was required to purify the tainted edifice. Special prayers over the new church included these phrases: "expel the Jewish error" and "clean this place of the stink of Jewish superstition."[109] Just as the case with Christian services held during Holy Week, Jews were sometimes assaulted as a result of the ceremony.

Perhaps the most famous incident of synagogue destruction and its political ramifications occurred in 388 when a synagogue in Callinicum (located on the banks of the Euphrates about 200 miles from the Mediterranean) was burned by a Christian crowd stimulated by the sermons and led by the local bishop and was transformed into a church. The Roman governor punished the arsonists and ordered the bishop to pay to rebuild the synagogue from his own funds, the decision being confirmed by the Emperor Theodosius I. It is clear that Theodosius had had no legal right to deny the Jews compensation for their loss, since they were still a legally recognized religion in the Empire.

At this point, however, Ambrose, the very influential bishop of Milan, wrote to the emperor defending the incendiaries and indicating that he would have loved to have done the burning himself.[110] Like Hilary of Poitiers, Ambrose saw any contact with Jews as defilement, for he associated the synagogue and the Jewish people with poisoned serpents and devils. In an exegesis, Ambrose had asked, "Wasn't it the Jewish people in the synagogue who are possessed by the unclean spirit of demons—as if bound fast by the coils of a serpent and caught in the snare of the devil—and who polluted its pretended bodily purity with the inner filth of its soul?"[111] The Gospel portion that stirred this commentary, Luke 4, told of Jesus' visit to the Capernaum synagogue where a man with "the spirit of an unclean demon" challenged him. (This followed Jesus' visit to the Nazareth synagogue where he first enunciated the purpose of his mission to preach about the Kingdom of God and where the congregation rejected him and allegedly tried to kill him.) Thus Luke came to the Middle Ages as interpreted by Ambrose in the work of Thomas Aquinas, who quotes Ambrose exactly in his *Exposito in Lucam*.[112]

Ambrose's argument in defense of the attack on the synagogue at Callinicum was that God had already destroyed the synagogue both as symbolic of Judaism in general and as a physical manifestation of local Jewish communities. Ambrose had already written elsewhere that for Christians to destroy a synagogue was a glorious act, so that "there might be no place where Christ is denied. [Because the synagogue is a] place of unbelief, a home of impiety, a refuge of insanity, damned by God himself."[113]

One Sunday, Ambrose preached a sermon on the Church and Synagogue attended by Emperor Theodosius, who had recently been excommunicated by him and was now repentant and very much open to his influence.[114] Face to face with the emperor, Ambrose reproached him for his action in support of the Jewish claims, arguing that it was a moral act to burn synagogues and if the laws forbade it, then the laws were wrong. Refusing him communion, he threatened that the emperor and his sons would be excommunicated again unless he rescinded his penalties against the incendiary bishop. In the end, Theodosius promised to do what Ambrose demanded.

This whole affair was of great importance, for here we have an example of how an assertive prelate, in whose mind even fairness to the Jews was incompatible with Christianity, could bully an emperor into an illegal action when it came to oppressing Jews. That is, it demonstrated what control the Church could wield over the secular authorities and their legislation on the Jews and that the Church's opinion could have an impact in the political world outside the Church. Granted, the Church was not omnipotent. In this particular case, its leverage was exhausted after five years when Theodosius and his sons reminded the Empire that the Jewish sect was legal and synagogue attacks "in the name of the Christian religion . . . must be suppressed" (*C.T.* 16.8.9).

John Chrysostom: Slay the Jews!

John Chrysostom was the Church Father most bitterly hostile to the Jews. This doctor of the Church, the greatest of the Greek Fathers, lived in Antioch, one of the Roman Empire's most cosmopolitan cities and the location where the disciples of Jesus were first called Christians.[115] During the life of this priest, patriarch, and preacher the ideological ascendancy of Christianity was not yet secure; Hellenism, Arianism, and Judaism still threatened.

Chrysostom's sermons addressed Christian Judaizers, since Christianity and Judaism were not yet absolutely distinct from one another, and Christians aware of the connections between the two religions were tempted to try blending them.[116] He resented his parishioners' observing the Sabbath, listening to rabbis' sermons and having rabbis bless their crops and visiting Jewish doctors for medical treatment. (In 300, the Church Council of Elvira threatened to excommunicate Christians who allowed their crops "to be blessed by the Jews, lest our benediction be rendered invalid and unprofitable.")[117] Choosing to discourage judaizing by damning Jews and Judaism, Chrysostom labeled Christian judaizers diseased enough or mad enough "to enter into fellowship with those who have committed outrages against God himself."[118] He sought to alienate any Christian feelings of affinity with, or common humanity toward, the Jews. The Jews were not ordinary members of the human race but people who "danced with the Devil."[119] Chrysostom also accused the Jews of being congenitally evil. They were evil before, during, and after the advent of Jesus, and their wickedness continued to his own day.[120] If a Jewish Scriptural passage referred to Jews as having worshipped idols, Chrysostom applied this prophetic self-criticism to his Jewish neighbors. In his exegesis on Psalm 106:37—"They sacrificed their sons and their daughters to the demons"—he wrote that the Jews had "slaughtered their progeny with their own hands to serve the accursed demons, who *are* the enemies of our life."[121]

Chrysostom believed that the Jews were deicides who had no chance for "atonement, excuse, or defense."[122] Jewish God-murder was "the crime of crimes"; Jews "crucified the Christ whom you [Christians] adore as God." (Chrysostom was perhaps the first to call the alleged Jewish crucifixion of Christ *theoktonian*, or deicide.)[123] Calling the Jews "the Synagogue of Christ-killers," he upbraided those Christians who had anything to do with "those who [nearly 400 years before] shouted, 'Crucify him. Crucify him.' "[124] He reports that a Jew of Antioch told him proudly, centuries after the event, "I crucified him."[125]

Chrysostom admitted that both testaments were the word of God. But he was especially angry that the Jews disregarded "the testamental references to Christ."[126] "I hate the synagogue and abhor it," he wrote, because the Jews "have the prophets but do not believe them; they read the sacred writings but reject their witness—and this is a mark of men guilty of the greatest outrage."[127] He saw the Jews as having "a veil lying over their hearts," which prevented them from understanding the christological meaning of their own Scriptures. "O what stupidity! what idiocy!" he exclaimed. Calling the Jews "ungrateful and unresponsive" and "shameless," he developed Paul's notion in Romans that only Christians are truly—spiritually—circumcised, whereas the blind Jews persist in regarding circumcision as physical.[128]

Chrysostom also paralleled Jerome's emphasis on the Jews' materialism. Gluttons, dull of mind, carnal of soul, "given to the things of this world," the Jews were spiritually blind to the truth of Christianity. Chrysostom accused the Jews of "plundering," "covetousness," "abandonment of the poor," "thefts," "cheating in trade." They "would dare anything for the sake of money," even the murder of God.[129]

Chrysostom found Christian reverence for the synagogue highly objectionable. The Godforsaken synagogues were worse than brothels and dangerous cliffs, more dangerous than dens of thieves and "filthy wild beasts," for synagogues seemed to be holy places. Christians, he implored, must continue to bypass Jewish places of worship, which he called "the Devil's house," as are "the souls of the Jews."[130]

> Here [in the synagogue] the slayers of Christ gather together, here the cross is driven out, here God is blasphemed, here the Father is ignored, here the Son is outraged, here the grace of the Spirit is rejected. Does not greater harm come from this place [than from pagan temples] since the Jews themselves are demons? . . . What worthy name can we find to call their synagogues? The temple was already a den of thieves. . . . Now you give it a name more worthy than it deserves if you call if a brothel, a stronghold of sin, a lodging-place for demons, a fortress of the devil, the destruction of the soul, the precipice and pit of all perdition . . . [131]

John's advice to Christians when happening upon a synagogue: "Must you not despise it, hold it in abomination, run away from it." [132]

Unlike Augustine, who found the Jews useful as exemplars of well-deserved suffering, Chrysostom denied any good use for Jews. Because "Jews rejected the rule of Christ," Chrysostom wanted these useless Jews killed. Just as animals that refuse to pull the plow are slaughtered, so Jews "grew fit for slaughter. This is why Christ said: 'As for these enemies of mine, who did not want me to reign over them, bring them here and slay them before me.' " Lest we miss his point about murdering the "useless" Jews, Chrysostom repeats it, adding a reference to Luke 19:27, which, he claims, refers specifically to a command of Jesus that the Jews be murdered.[133] Chrysostom later justified such an atrocity by arguing that "what is done in accordance with God's will is the best of all things even if it seems bad. . . . Suppose someone slays another in accordance with God's will. This slaying is better than any loving-kindness."[134]

John Chrysostom was an enormously influential preacher. Hitler expressed his admiration for the anti-Jewish ideas of "all genuine Christians of outstanding calibre," among whom he counted John Chrysostom. "The Jew regards work as the means to the exploitation of other peoples," Hitler noted. "The Jew never works as a productive creator without the great aim of becoming the master. He works unproductively, using and enjoying other people's work." Later in the same speech, Hitler discussed the Gospel's picture of Jesus and the Jews: "My Lord and Savior . . . recognized these Jews for what they were and summoned men to the fight against them . . . ."[135]John's homilies are found in nearly 200 collections of his work in the ancient world and excerpts from them were inserted into the

Byzantine Holy Week liturgy. Although a few comments of his indicate a sneaking respect for Jews, his major contribution to Jewish–Christian relations was an encouragement of violence against the Jewish people and their synagogues.[136] His anti-Jewish diatribes seem to have had no immediate effect, but riots occurred 28 years later against the synagogues of Antioch. The attack on these houses of worship was sparked by a rumor that a Christian boy in a nearby town had been murdered by Jews. Were Christians not more likely to believe such rumors having been exposed to Chrysostom's hatred? Chrysostom's martyrdom in 407 lent prestige to his oral and written assaults on the Jews and influenced Christian attitudes toward their Jewish neighbors. Chrysostom's attitudes may also have influenced the anti-Jewish actions of Byzantine emperors. His anti-Jewish sermons were translated into Russian at a time when, in 1100, the first pogroms in Russia were taking place.[137] From the eighth century on, Chrysostom's homilies against the Jews were taught to priests in Christian schools and seminaries.[138]

Jewish religious, economic, legal, and political privileges in the Western empire ultimately depended on Christian theological premises, for the Christian-imperial laws embodied in the Theodosian Code determined what the Jews could or could not do. The emperors issued the laws; the Church conditioned the emperors; and the theology of triumph determined the Church's Jewish policy. The Church and the Christian emperors made Judaism a crime.[139] The way was left open for the legal destruction of synagogues, forced baptisms, and the medieval trials and burnings of the Talmud, as well as medieval defamations and massacres.

Similar injustices and persecutions took place in the Byzantine Empire, founded in 330 and lasting until Constantinople's capture by the Turks in 1453. The millions of Jews of the East presented a significant challenge to Christian thinkers, the state, and society. (The Jewish population dropped from perhaps six million in the first century C.E. to only 100,000 in the twelfth.)[140] Eastern theologians replicated the anti-Jewish theology of their Western colleagues, and the Eastern empire responded in the same fashion as in the Western empire. Before Justinian's reign, 527–65, the Jews were tolerated but discriminated against. As J.F. Haldon has written, the Jews were "the one group of believers who presented a real problem for both Christian theologians and the Christian state throughout . . . the history of the [Byzantine] state." Although the Jews managed to survive and sometimes to prosper, they were nevertheless subject to "constant persecutions, occupying always a subordinate position within Eastern Christian society and culture."[141] Eastern emperors promulgated several laws that undermined Judaism economically, politically, and religiously.

In 396–7, Emperor Arcadius confirmed Jewish privileges. Arcadius issued laws that prevented outsiders from determining the prices of Jewish merchandise, that provided punishments for insults to the Jewish patriarch, that protected Jewish synagogues against assault, and that reaffirmed traditional Jewish rights.[142] But when John Chrysostom became Christian patriarch in 399, Arcadius repealed the exemptions of Jews from performing civic duties and prohibited the Jewish patriarch from collecting taxes.[143] John Chrysostom's attitudes must have been a factor in Arcadius' changed Jewish policy, because when Chrysostom was expelled from Constantinople in 404, Arcadius reestablished a policy favorable to the Jews and reasserted the privileges of the Jewish patriarch.[144] One law specified, "We have formerly ordered that what was customarily contributed to the patriarchs by the Jews . . . shall not be contributed at all. Now, however, with that order revoked, we want all to know that our clemency has granted to the Jews the rights of conveyance according to the privileges established by the ancient emperors."[145]

Arcadian leniency toward the Jews was, however, often replaced by persecution, which was seldom as extreme as the Western model.[146] Without any inherent rights, the Jews were always subject to potential destruction. A long series of persecutions starting in the fifth century resulted in an ineffective attempt to convert Jewish residents, which was repeated sporadically and unsuccessfully throughout Byzantine history.[147] The forced baptisms were ineffectual, since the Jews, following an often repeated pattern, "submitted to baptism and then washed themsevles clean of it . . . and thus contaminated the faith."[148] The Emperor Zeno (d. 91) asked at the burning of Jewish corpses in the Antioch synagogue in 489, "Why did they not burn the living Jews along with the dead? And then the affair would be over."[149] In the sixth century, both Justin I and Justinian I began to apply laws against heretics and pagans to Jews. The anti-Jewish policies of Byzantine rulers resulted in the Jews being deprived of the ability to serve the state, make wills, receive inheritances, participate in legal actions, or serve as witnesses in court.

Justinian's collection of laws concerning the Jews (promulgated between 529 and 534) reflected more clearly than Theodosius' code the influence of Christian anti-Jewish precepts. Justinian's more anti-Jewish version of Roman law became the standard basis of medieval law from the twelfth century forward, succeeding the less oppressive version of the Theodosian Code, the Arian-Visigothic *Breviarium Alaricianum*.[150] Although Judaism was the acknowledged root religion of Christianity, Justinian confirmed the Theodosian code's reduction of the Jewish status to "permanent inferiority."[151] He dropped several Theodosian laws legitimizing Judaism, establishing Jewish privileges, and protecting Jews.[152]

Judaism was no longer a *religio licita*. Based on Christian tenets and the Church's authority, some of Justinian's first laws removed Jewish rights and classified Jews, along with other non-Christians, as heretics.[153] Later laws, published "in the name of our Lord and God Jesus Christ," punished Jewish marriages as "abominable," forbade circumcision of converts, and limited the way Jews could worship. "It is totally absurd to permit wicked men to deal with sacred matters."[154] Justinian formulated these laws in language taken directly from John Chrysostom, and invited the ecclesiastical authorities to oversee the Jews' status and rights.[155]

The Justinian Code also attempted to determine the tenets of Jewish belief. Based on the ideas of Jerome and other fathers of the Church, Justinian's *Novella* 146 stated that the Jewish Mishnah intentionally distorted the authentic "christological" meanings of the Jewish Testament, and therefore it should be forbidden to the Jews. The same law stated that Jews should interpret their holy books only as "announcing the Great God and the Savior of the human race, Jesus Christ," not for their "literal" meaning. Rabbinic interpretations were not permitted. Jewish exegesis was "malignant," "extraneous and unwritten . . . ungodly nonsense." Jews who denied the Resurrection or the Last Judgment would suffer "the harshest punishments."[156]

The Byzantine Jews experienced the same kind of inconsistent treatment that the Jews in Western Europe encountered during the early Middle Ages. Long periods of benign neglect alternated with open persecutions whose origins lay in long-term Christian theological and social antagonisms and short-term internal and external stresses in the empire. Several Byzantine emperors unsuccessfully attempted to convert Jews forcibly: Maurice (d. 602), Phocas (d. 610), Heraklios (d. 641),[157] Leo III (d. 741),[158] Basil I (d. 886), Romanos I Lekapenos (944), and later John Vatatzes (d. 1254).[159] As a result, many Jews fled to the Khazar kingdom. Although the Orthodox Church officially opposed forced conversions, its theology, propaganda, and liturgy supported imperial hostility toward Jews.[160]

Despite their theological alienation, Byzantine Jews partook in the dominant Greco-Roman culture. Although as Jews they suffered the restrictions of second-class citizenship, they were the only non-Christian ethnic group that the empire tolerated. [161] In the ninth century, several iconoclastic emperors were sympathetic toward the Jews for reasons of state. Michael II (d. 829) so favored them that he was accused of judaizing; Leo VI condemned anti-Jewish violence and established a secure basis for the existence of the Jewish community.[162]

Once into the eleventh century, the Byzantine Jews suffered disdain but no catastrophes. Byzantine Jews were permitted to engage in a variety of occupations, not just moneylending. Whereas Latin-Christian Crusaders assaulted, brutalized, and murdered Jews, Byzantine Christians and Jews joined to defend themselves against the Western "soldiers of Christ." Eastern-Christian and Jewish scholars associated more frequently than their Western counterparts and Eastern theologians were more moderate in their opposition to Jews. At the end of the twelfth century, Benjamin of Tudela reported that Christians hated and oppressed the Jews of Constantinople, that the Greeks defiled the Jews by pouring out their filthy water in the ghetto and "beat them in the streets."[163] But, under the protection of the king's Jewish physician, Solomon ha-Mizri, "the Jews are rich and good, kindly and charitable, and cheerfully bear the burden of their oppression."[164]

In 1259, a new dynasty took power with Michael Palaiologos. He revoked his predecessor's order of forced baptism and awarded Jews religious and economic liberties in return for their backing. Michael needed Jewish support because his economically and militarily exploited Orthodox subjects opposed him. This more tolerant imperial policy persisted until the fall of Byzantium to the Turks in the mid-fifteenth century.[165]

Despite the more benign Christian treatment of Byzantine Jews, especially after the twelfth century, the Orthodox Church consistently taught Christianity's triumph and Judaism's inferiority. The West portrayed synagoga as corrupted by the devil and damned to hell by Christ, but in the East, synagoga was simply rejected in favor of ecclesia.[166] But other times the Orthodox Church pictured Jews as on the wrong side in the metaphysical battle between the angels and devils. Because Jews represented the most attractive alternative to Orthodoxy within the empire, since Judaism was based on the same bible as Christianity, the Orthodox Church insisted that the Jews be degraded. Orthodox Christians employed well-rehearsed anti-Jewish arguments against heretics (Latin Christians, Moslems, and Bogomils) and even labeled them "Jews."[167] Some Orthodox clerics argued that Jews who did not accept Jesus as the Christ were not authentic Jews, since they did not follow Moses through whom God had predicted Jesus' coming as Messiah.[168]

Although at a lower level of antagonism, the Eastern Church nevertheless held Jews in the same contempt as the Western Church. The Jews of Corfu were stoned during Easter, and the Cyprus passion cycle degraded Jews.[169] In 1200, a visiting Russian prelate, Anthony, archbishop of Novgorod, was informed that "a Jew had struck the neck of [an icon of "the Holy Virgin holding the Christ"] with a knife and it issued blood." The same prelate prayed that God "will lead the accursed Jews to baptism."[170] By the early fourteenth century, although the Byzantine state tolerated Jews, the Eastern Church grew more antagonistic toward them. Urging the autocrator to evict the deicidal Jews from Constantinople, the Patriarch Athanasios called the Jews "the deicidal Synagoga"[171] and asserted that "[it is the duty of Christians] to hate the deicidal Jews."[172] About 1300, a monk, Maximus Planudes, citing John the Baptist, referred to the Jews as a "brood of vipers" and to the stink of Jewish belief as disgusting as "the foul smell from their tannery."[173] The Greek monk Matthew Blastares wrote the *Syntagma* (1335), an encyclopedia of earlier secular and ecclesiastical rulings concerning the Jews. It repeated traditional Latin Christian anti-Jewish attitudes, calling Jews "stiff-necked" and "uncircumcised in the heart." Even before the killing of Christ, God "detested" the Jewish rites. Christians who participate in these rites, use Jewish doctors, bathe with Jews, or associate with Jews in any friendly way are to be excommunicated.[174]

Patristic theologians regarded the Jews' habits and institutions as hateful indications of the sinfulness and criminality of a Judaism so deaf and blind to Christian truth and goodness that it failed to disappear at the coming of Christ. This message, imbedded in the Christian mind and institutionalized in the triumphant Church and Christianized law, had a devastating effect on the long-term status of Jews in Christian Europe.

When these theological writers attacked Jews, they drew little, if any, material from contemporary Jewish behavior; the phrases "a Jew" and "some Jews" are nearly unknown in patristic writings. Instead they implicated all Jews in mythic religious crimes, in a continuous history of "incurable vices," which climaxed in the gravest possible crime and sin, the murder of God. The catalogue of Jewish "crimes" was endless. Gregory, the fourth-century bishop of Nyssa described the Jews as:

> "Murderers of the Lord, killers of the prophets, enemies and slanderers of God; violators of the law, adversaries of grace, aliens to the faith of their fathers, advocates of the devil, progeny of poison snakes, . . . whose minds are held in darkness, filled with the anger of the Pharisees, a sanhedrin of satans. Criminals, degenerates, . . . enemies of all that is decent and beautiful. They are guilty of shouting: Away with him, away with him. Crucify him. He who was God in the flesh!"[175]

To both Western and Eastern Christian theologians of the Common Era's first five centuries, Jews were devilish Cains, Judases, and Antichrists.[176] Since Jews were inherently evil, since they enticed Gentiles away from the true faith and lied about Christianity, since they tempted Christians to sin and obstructed divine teaching, since they murdered Jesus and rebelled against the coming of the Kingdom of God, the Jews were the moral equivalent of the devil.[177] Coupling the devil and the Jews, Eusebius of Alexandria began *every* paragraph in the first half of his sermon on the

Resurrection in this way:

> Woe to you wretches, . . . you were called sons and became dogs. Woe to you, stiff-necked and uncircumcised [Jews were commonly accused of being circumcised only in the flesh, but not in the spirit], from being the Elect of God you became wolves, and sharpened your teeth upon the Lamb of God. You are estranged from His Glory; woe to you, ungrateful wretches, who have loved Hell and its eternal fires. . . . Hell . . . shall imprison you with your father the devil.[178]

For many Christians, attitudes favorable to Jews "were lost in the blood of Calvary"?[179] But deicide itself did not make the Jews evil. It was the Jews' evil that caused them to commit deicide and myriad other crimes.[180] Just as the Christian Scriptures provided a sacramental basis for the patristic interpretation of Jewishness, so the Church Fathers provided consecrated ammunition for contemporary and future assaults on Jews and Judaism. (It has been estimated that medieval anti-Jewish writings amount to more than 1,000; Jewish anti-Christian writings amount to about 250.)[181]

As both Jews and Christians moved to the end of the first and the beginning of the second millennium of the Common Era, the average Christian began to respond more than ever before to the anti-Jewish propaganda of the Latin Church based on its *theologia gloriae*, embodied in its consiliar proclamations, papal bulls, creeds, and liturgy, and dominating the law of nearly all of Christian Europe. Yet the Jewish spirit stubbornly persisted as a living challenge to Christian identity, just as it had for a thousand years. Judaism continued to be Christianity's most formidable ideological opponent.[182] To combat this challenge, the theological propagandists of the Church made sure that the Jews came to be considered the worst enemies of humanity, and almost any harm done to them was considered, if not a blessing, at least justified by their "continuing crimes" of deicide, blasphemy, heresy. More than ever before, Jewish lives were in the hands of their worst enemies.

Jewish Self-Definition

In the earliest centuries of the Common Era, Jewish responses to developing Christianity were remarkably mild.[182] Surprisingly, the Mishnah was silent on Christianity, even though it was composed when the Christian theologians of the late Roman and early medieval periods were engaged in the most hateful and hurtful polemic against all things Jewish.[184] The Jerusalem Talmud, an elaboration of the Mishnah, however, reflects a defensive Jewish reaction to Christianity: "If someone [Jesus] will tell you, 'I am God,' he is a liar; 'I am son of man,' his end is that he will regret it; 'I am going to Heaven,' he says this but will not fulfill it."[185] The wonder was that the Jews did not spend more time and effort on religious condemnation of Christianity.[186] The Talmud devotes no more than twenty pages out of thousands to Jesus and Christianity.

But the more Christianity appropriated Jewish traditions, Christian apologists increased their attacks, and the Christianized Roman empire added discriminatory laws, the more the Jews sought to preserve their own identity. One means of achieving this goal was to keep heretics, informers, Christians, and Jewish-Christian apostates out of synagogues.[187]

In the first two centuries following the destruction of the Temple in 70 C.E., Christianity had seemed to be an internal Jewish movement intent on converting the traditional synagogue. In order to distinguish Jews from Christians of Jewish background, and exclude the latter from the synagogue itself, the Jewish Patriarch Rabbi Gamaliel II ordered that an alternative form of the *Amidah*, or *Shemoneh Esreh* (*The Eighteen Blessings*), be included in the Jewish liturgy.[188] The first-century twelfth benediction, the *Birkat ha-Minim*, the "*Blessing" of the Heretics*, included in the *Shemoneh Esreh*, was originally intended simply to prevent heretics and Jewish-Christians from participating in the synagogue service, not to curse them or to expel them from Judaism. It read as follows:

> For the apostates may there be no hope unless they return to your Torah. As for the *Noserim* [Gentile Christians?] and the *Minim* [Jewish-Christians?], may they perish immediately. Speedily may they be erased from the Book of Life and may they not be registered among the righteous. Blessed are You, O Lord, Who subdue the wicked.[189]

Christian writers also refer to the *Shemoneh Esreh*. The Gospel of John refers three times to the expulsion of Jewish-Christians from the synagogue, which may have been as a result of the benediction against the *minim*. In the middle of the second century, in his *Dialogue with Trypho*, Justin Martyr accused the Jews of "cursing in your synagogues those that believe in Christ." In the third to fourth centuries, Origen and Epiphanius also mention the thrice-daily Jewish curse against Christians. Early in the fifth century, Jerome claimed that "three times every day in the synagogues they anathematize the appellation 'Christian' under the name of 'Nazarenes.' "[190]

The Gospels and Pauline epistles often quoted from the Torah as a means of justifying, sanctifying, and spreading Christianity; the rabbis insisted that these Christian texts were without validity, sanctity, and halachic status. The Talmud (*Shabbath* 116a) also argues that the books of the *Minim* (Christian Scriptures) contain such blasphemy that they should not be rescued from fire, water, or a cave-in.[191] For the *Minim* know God yet deny God. They "stir up jealousy, enmity, and wrath between Israel and their Father in Heaven"; they also lead others from God.[192]

Four categories of unfaithful Jews were *Minim* (Jews false at heart to Judaism), *Meshummadim* (apostates who willfully transgress Jewish ceremonial law), *Masoroth* (political betrayers), and *Epiqurosin* (free-thinkers). The *Minim*—Jewish hypocrites who "did not walk after their hearts"—were dangerous because they were secret apostates who did not admit their apostasy; they had to be discovered and expelled from Judaism.[193] *Minim* may have referred to Jewish-Christians; in contrast to *Noserim*, or Nazoreans, which may have applied to Gentile Christians. Marcel Simon and Lawrence Schiffman suggest that the term *Minim* in this context refers originally to Jewish-Christians and in the second or third centuries to Gentile Christians as well.[194] Reuven Kimelman, however, argues that the anti-*Minim* prayer never applied to Christians of Gentile background, who, based on the complaints of the Church Fathers Origen, Jerome, and John Chrysostom, were evidently welcomed into the synagogue. The rabbis wanted to expel only Jewish sectarians, the Nazoraeans, from the Jewish community.[195] Either no single Jewish approach to relations with Gentile Christians existed or the evidence of the Church Fathers was not accurate.

Despite Talmudic sayings accusing Jesus of being a Jewish fraud dating from at least the fifth century and despite the Christian attack on Judaism, Jewish thought provided for the salvation of non-Jews, or Gentiles.[196] In contrast to the Christian damnation of non-Christians, the Talmud states, "Among all the nations there are just individuals, and they will have a share in the world to come."[197] The Laws of Noah specify that Gentiles would be considered "the pious ones among the nations of the world deserving a share in the world to come," if they were not idolaters, blasphemers, immoralists, murderers, or thieves; not cruel and unjust; and if they were righteous.[198]

Jewish theology and Christian theology also differ concerning the treatment of the stranger in the midst of the faithful. Whereas Christianity maintained that there could be no salvation without membership in the Church, Jewish Scripture mandated that the stranger be protected. Leviticus states that "when an alien resides with you in your land, you shall not oppress the alien. The stranger shall be to you as the citizen among you; you shall love this stranger as yourself, for you also were aliens in the land of Egypt."[199] The Jewish Scriptures also contain entire books written in praise of non-Jews, such as Job and Ruth, who are portrayed as walking in righteousness before God. Ruth's son, Obed, became the grandfather of David, one of the most illustrious sons of Israel.[200] In the Middle Ages, Maimonides stated that all Gentiles who bring their souls "to perfection through virtues and wisdom in the knowledge of God [have] a share in eternal blessedness."[201] The Jews' idea of "the nations" allowed Christians and other peoples to coexist with Jews so long as they allowed the Jews to be themselves.[202]

The rabbis criticized christological arguments and emphasized that only they understood God's teachings, or Torah, in learned disputations with Christian theologians.[203] Jerome described his rabbinic opponents as having "loose lips and twisted tongues, with spit spluttering from their shaven, grinning jaws."[204] In reaction to the Christian Fathers' claim that only Christians were the true Israel, the Jews developed the doctrine of the Dual Torah: God was revealed in the Torah given to Moses at Sinai but the Torah was expressed in two media: the written Torah and the oral interpretation of these writings.[205] The oral revelation, that is, the teachings of the sages, was systematized and written down first in the Mishnah[206] and later in the Talmud between the third and fifth centuries C.E.[207] The Talmud stated, "God anticipated that the gentiles would translate the Torah, read it in Greek, and say: 'We are Israel, and we are the children of God.' [But those] who possess My secret writings are My children. And what are these writings?—the Mishnah."[208]

Chapter 3

Medieval Violence

As soon as a layman hears a Jew malign the Christian faith, the Christian should defend it by striking the slandering evil-doer with a good sword thrust into the belly as far as the sword will go.

—*St. Louis*

In his great work, *A Social and Religious History of the Jews*, Salo Baron objected to the "lachrymose" interpretation of Jewish history.[1] Surely, tears can distort any historian's vision. But when all is said and done, after we take full cognizance of the positive characteristics of the indomitable Jewish spirit, after we appreciate the decent relationships that have always existed between some Christians and Jews, we must nevertheless admit that the written evidence clearly indicates that the predominant experience of Jews in Christendom has been one of, in the words of the sixteenth-century Italian-Jewish chronicler Joseph ha-Cohen, "a vale of tears."[2]

During the Middle Ages, as R.W. Southern has observed, "Church and society were one . . . . This is the clue to a large part of European history, whether secular or ecclesiastical."[3] During this period, the fundamental Church policy was to degrade Jewish life and alienate Jews from Christian society to prove that God had rejected the Jews and had chosen the Christians to be God's people. The Jewish foundation of Christianity was seldom acknowledged; instead, the religion of the Jews was cast in an evil shadow, ridiculed, tried, and convicted of blasphemy. Although the Church often sought to preserve Jewish lives—the idea was to make Jews suffer, according to the authoritative Augustine, not kill them—many Christians, influenced by the Church's overt hostility toward Jews as "negators and malefactors," murdered Jews, often in large numbers.

Christians did not perceive the realities of contemporary Jewish life. They did not see the Jews for what they were, a hardworking, sometimes desperate, minority cleaving with all its energy to its religious and community values in a hostile Christian sea. Instead, they viewed Jews darkly through a theological lens as Christendom's greatest sinners and enemies, as beasts, as the very model of evil.[4] In the Dark Ages, Jews were known generally as *mercatorii* but in the (High) Middle Ages as *usurarii*.

The humane relationships between Jews and Christians in the Early Middle Ages caused Church officials, fearing lest the contemptible Jews corrupt Christians, to work all the harder to alienate the two groups. In twelfth-century England, the Jew Benedict of York had been baptized under the threat of death but then, as a "convert without conviction,"[5] he recanted before King Richard I. Baldwin, the archbishop of Canterbury, then burst out in anger: "Since he does not wish to be a Christian, let him be the Devil's man." Benedict had, the chronicler wrote, "returned to the Jewish depravity . . . like a dog to his vomit." (The Visigothic Profession also stated: "I will never return to the vomit of Jewish superstition, . . . or [associate] with the wicked Jews." The Council of Agde (506), the Crusader chronicler Ekkehard of Aura, and Alexander of Hales, a thirteenth-century canonist used the same image. Pope Benedict XII wrote in 1338 about a relapsed convert from Judaism "returning to the Jewish error like a dog to its vomit.")[6] In England and France, Christians who had sex with or married a Jewish woman were accused of bestiality; the legal opinion was that "coition with a Jewess is precisely the same as if a man should copulate with a dog."[7] The Council of Exeter decreed in 1287 that Christian women were not to serve in Jewish homes, because "consorting with evil corrupts the good."[8] Christians imagined the existence of a satanic trinity, consisting of the devil, the Antichrist, and the Jew, antithetical to the Holy Trinity.[9]

Centuries of anti-Jewish propaganda had so confused average Christians that the Jew they met on the streets became identified with the Jew-Devil of Christian myth. In a violent, superstitious, and emotional period of history, the Church's anti-Jewish doctrines—although often complicated by economic and political considerations—comprised "a standing incitation to maltreatment and murder."[10] If Gavin Langmuir and David Berger are correct, Christians during this period developed doubts about the validity of their own Christian beliefs. Men such as Peter the Venerable and Peter Damian hated the Jews because they themselves were profoundly disturbed by the conflict between historical realities and theological perceptions.[11] Norman Cohn suggests that European Christians found Christian moral teachings impossibly demanding and therefore these medieval Christians felt an "unacknowledged hostility to Christianity." Their unconscious anger was expressed by attacking as devilish those out-groups that opposed the

Christian faith, such as heretics, witches, and Jews.[12] Especially once the Inquisition was established, at the turn of the twelfth century, Jews were treated not as members of an independent faith but as perverse deviants from the one true faith. They "knew the truth and rejected it."[13]

These ideas and images of the Jew provided a moral and intellectual pretext for the princes and people when they mistreated their Jewish subjects and neighbors. As a result, the Jewish communities of Christian Europe grew ever more vulnerable to violence. Jewish suffering was seen no longer as some hypothetical divine judgment that had been meted out to the Jews of the past. Instead, people were convinced that contemporary Jews should suffer for the eternal sins of the Jewish people. Therefore, many Christians decided that they would become the executioners of God's condemnatory judgment through discriminating against, expropriating, expelling, and mass murdering Jews.

The intensity of the Christian assault on Jewishness was also aggravated by the millennial Christian belief that all of Christendom was a holy *corpus mysticum christi* (the mystical body of Christ) without a legitimate place for the Jews. This conception of Christian society as a mystical body of Christ that totally excluded Jews led to attempts to convert them, expel them, or exterminate them. Through clerical sermons resounding with theological anti-Judaism, several friars of the newly founded mendicant orders directly or indirectly encouraged the faithful to believe anti-Jewish defamatory myths and therefore to do violence to Jews. In the context of endemic socioeconomic crisis, ecclesiastical divisiveness, wars, famines, and plagues, standard Christian beliefs were not emotionally satisfying, and radical doubt about the validity of Christianity often crept in to challenge the religious self-confidence of the Christian populace. This in turn led to what Gavin Langmuir calls "delusions" about the Jews,[14] namely, the false beliefs that they committed abominable crimes, not only by murdering the Christ, but also by assassinating contemporary Christians; they paralleled Hitler's fantasies. (Pagans Democritus and Apion accused Jews of ritual-murder, John Chrysostom hinted at it when he wrote that the Jews had "slaughtered their progeny with their own hands to serve the accursed demons, who *are* the enemies of our life,"[15] and Christian charges of deicide, devilishness, betrayal, and worship of money occurred already in the first centuries of the Christian era.) As a result, Jews came to be considered Christendom's most flagrant public enemies.[16] Anti-Jewish propaganda became more virulent and far more effective, the Church allowing the Jews no inherent right to exist as a free people among Christians.[17] Although as a medieval corporation the Jews sometimes enjoyed a degree of self-government, the Jews' right to exist at all was oftentimes challenged.

Jews were often degraded, usually landless, politically powerless, frequently dishonored, and denied any inherent right to exist in Christian Europe; they lived there on the sufferance of popes and Christian princes. The Jews were sometimes treated as a privileged group and often lived better than serfs. But a privilege depends entirely on the power and the whim of the grantor.[18] Once the Jews' utility to the princes diminished—often because of ecclesiastical pressure—the Jews' lives and property were often forfeit. The Church had a theological and psychological need for Jewish status to be lower than that of any Christian.

By the twelfth century, the status of the Jews in Christian Europe dropped below that of all Christians. Bernard of Clairvaux, the greatest spiritual figure of the twelfth century, observed: "there is no more dishonorable or serious serfdom than that of the Jews. They carry it with them wherever they go, and everywhere they find their masters."[19] According to canon 24 of the Third Lateran Council of 1179, "Jews should be slaves to Christians and at the same time treated kindly due of humanitarian considerations."[20] Canon 26 held that "the testimony of Christians against Jews is to be preferred in all causes where they use their own witnesses against Christians. And we decree that those are to be anathematized whosoever prefer Jews to Christians in this regard, for they ought to be under Christians . . . ."[21] In the 1180s the French king Philip Augustus seized the Jews' property and then expelled them from his domain, without troubling to find legal justification. As Langmuir noted, "the king never could have treated his serfs this way and never did so."[22]

The most influential of medieval theologians, the thirteenth-century Dominican Thomas Aquinas, held that the Jews were slaves to the Church and to the Christian princes.[23] "The Jews are themselves the subjects of the church," wrote Aquinas, "and she can dispose of their possessions as do secular princes . . . . The Jews are the slaves [or serfs] of kings and princes."[24] His perspective greatly influenced the mendicant orders, the Inquisition, and the entire Church ever since. Although he felt that the Jews could keep property sufficient to their survival, he wrote to the duchess of Brabant that "Jews, in consequence of their sins, are or were destined to perpetual slavery," and as a result the princes can treat Jewish property as their own.[25]

No pope would warn *Christian* peasants that their very right to exist depended on his feelings of kindness; in his mandate to the archbishops of Sens and Paris in July 1205, however, Innocent III noted that only because of "kind[ness] we have tolerated these Jews."[26] Augustine's theological doctrine of the "Witness People," repeated by most medieval theologians,[27] gave medieval Jews at best a precarious protection, but, as the popes defined the piety and kindness of the Church, it provided Jews only the flimsiest of shields. These two principles, Christian kindness and Augustine's Witness

People doctrine, were the unsteady ideological bases for Jewish existence in Christendom.[28]

Deteriorated Jewish Condition

1. *Picture* and *word* elaborated the intimate relationship between the Jews and "the powers of evil": in Christian art, illustrated Bibles, prayer-books, and psalters; in statues and bas reliefs both inside and outside of churches and alongside public ways; on portals and walls, in tapestries, stained-glass windows, and furniture; on stoves and plates, bric-a-brac, and even on urinals.[29] In one sculpture and drawing, the Archangel Michael cuts the hands from a rabbi who dared attack Mary's funeral procession; the devil shoots Synagoga in the eye; Jesus himself assaults Synagoga, sometimes pushing her into Hell. Other times, the violence was projected onto the Jews themselves. One of the most popular motifs in Church art was the stoning of Stephen, based on the passage in Acts 7. In the twelfth-century relief on the cathedral at Rouen, the Jews are shown murdering Stephen just after he accuses them of being idolaters, "forever opposing the Holy Spirit," and murdering all the prophets as well as Christ.[30] In a thirteenth-century illuminated French manuscript, contemporary Jews were shown as hating Jesus so much that they stabbed his portrait in an assassination attempt, a clear iteration of the Passion.[31] Other illustrations in effect accused Jews of burning or stabbing the consecrated Host or murdering Christian children as further evidence of their continuing maliciousness.[32] In the fifteenth century, an illustrated Alsatian Bible represented Jewish children as passing stones to adult Jews who were trying to kill Jesus after he had declared himself Son of God.

2. Latin and vernacular *literature* reflected and endorsed theological defamations of Jews.[33] Medieval *theater* was extraordinarily popular—with morality, miracle, mystery, passion, resurrection, redemption, and *Corpus Christi* plays, as well as the *planctus Mariae*, or laments of Mary.[34] It reflects "a mood of unrelenting hostility . . . directed in large part toward the disbeliever, the Jew." Even when no Jews lived in the local area or within the whole nation, the Jew was denigrated.

3. By the end of the twelfth century, the walled *ghetto* marked most of Christian Europe's towns.[35] Although it may have afforded Jews some protection, the ghetto not only kept Jews and Christians apart, but it also maintained Jews in their subordinate and degraded status.

4. The Church created *stigmatic emblems* and ordered secular authorities to impose them in order to mark Jews as pariahs and isolate them from

Christians. The Fourth Lateran Council of 1215, after declaring that Jews would roast in hell for all eternity (at least until the Second Judgment), accused the Jewish "blasphemers of Christ" of "treachery" and "cruel oppression" in charging interest on loans to Christians and decreed that Jews must wear clothing that distinguished them from Christians because "through error [male] Christians have relations with the women of Jews . . . , and [male] Jews . . . with Christian women . . . Jews . . . of both sexes in every Christian province and at all times shall be marked off in the eyes of the public from other peoples through the character of their dress."[36] The canons of the Church Council at Basle, 1431–43, decreed the same kinds of restrictions, adding compulsory sermons. These canons were not rescinded until 1846.[37]

5. The Church's anti-Jewish theology served as the basis for its *magisterium*, or teaching, which in turn was embodied in *legislation*, ecclesiastical, such as papal decree and canon law, and, inevitably the Church's law influenced the laws of the land, the secular law codes, such as the *Siete Partidas* of 1263,[38] the very influential south German *Schwabenspiegel*, and the Viennese *Stadtrechtsbuch* were two other secular law codes based on Church teachings.

6. For much of the High Middle Ages in most of Europe, the Jews were marked out by *special commercial legislation* and severely restricted in their possibilities for making a living.[39] And so they resorted to usury. The Jews were left little choice, for they lived in scattered communities, and they suffered frequent expropriations, periodic expulsions, and persistent heavy taxation: they were taxed in their coming and going, in their buying and selling, in their praying and marrying, in their birthing and dying. They were frequently barred from owning land, from hiring non-Jewish workers, and from joining Christian craft and merchant guilds.[40] And as moneylenders, they were hated. Who better than the "Christ-killing, traitorous, malicious Jews" to handle the necessary but "filthy" business of lending at interest?[41] By the twelfth century, "Jew" and "usurer" became nearly synonymous all across Christian Europe.[42]

7. Many Christian theologians argued that the root of the Jews' refusal to accept Jesus as the Christ was the *Talmud*. The Talmud was also attacked by the Churchmen of the thirteenth century in France and later in Spain, Germany, and Italy, where it was tried and convicted of containing heresy.[43] Nicholas Donin, a Jewish convert to Catholicism who criticized the Franciscan Order and in 1287 was condemned to death by Pope Nicholas III, had denounced the Talmud to Pope Gregory IX who in turn in 1239 had ordered all prelates and princes of France, England, Spain, and Portugal to investigate the Talmud for blasphemy and heresy.[44]

Odo of Chateauroux, papal legate to the king of France, headed a papal commission evaluating the Talmud.[45] He concluded that the Talmud "contained so many unspeakable insults that it arouses shame in those who read it, and horror in those who hear it." Containing "errors, abuses, blasphemies, and wickedness" and injurious to the Christian faith, these talmudic books are essentially heretical and deserve to be burned. The eminent Franciscan biblical scholar, Nicholas of Lyra (d. 1340), agreed that traditional rabbinic Judaism as well as its contemporary representatives had maliciously deviated from Jewish Scripture and were essentially heretical.[46]

Louis IX (Saint Louis), king of France, commanded an investigation of the Talmud at Paris in 1240.[47] He argued that the Jews "polluted [France] with their filth."[48] His entourage was filled with clerics, in particular Dominicans and Franciscans.[49] Louis related to his devoted chronicler John of Joinville "that a layman, as soon as he hears the Christian faith maligned, should defend it only by the sword, with a good thrust in the [Jew's] belly, as far as the sword will go."[50] These attitudes may have encouraged the Church to canonize him in 1297 for his "Christian holiness."

Convicted of heresy and blasphemy against Mary and Christ, thousands of copies of the Talmud were burned.[51] The Talmud played such an essential role in Jewish culture and self-identity that once the Talmud was declared blasphemous, no medieval Jew could easily avoid the charge of defamation of Christianity.[52]

8. Jews were also identified with *sorcery, magic*, and *witches*.[53] This association of malevolents led to the word "synagogue" being used to describe an assembly of heretics or witches, where the devil was worshipped.[54] At the end of the sixteenth century, the Catholic elector of Trier and the Protestant duke of Brunswick set out to exterminate both witches and Jews. In 1609 French King Henry IV commissioned Pierre de l'Ancre as grand inquisitor of witches. This "bigoted Catholic" was a "gleeful executioner" who, writes Trevor-Roper, "gloried in his Jesuit education."[55] De l'Ancre saw Jews as "blasphemers . . . just another kind of witch."[56] He attacked Judaism as irrational and indecent; thought the Jews could turn into wolves at night; disdained their cruelty and rapaciousness, their murder of Christians by poisoning of the wells, and their forced circumcision and ritual murder of Christian children.[57] Perhaps worst, he observed that Jews "were more perfidious and faithless than demons." He concluded, "The Jews deserve every execration, and as destroyers of all divine and human majesty, they merit the greatest torment. Slow fire, melted lead, boiling oil, pitch, wax, and sulfur mixed together would not make tortures fitting, painful, and cruel enough for the punishment of such great and horrible crimes as these people commonly commit . . . ."[58]

9. Jews were forced to take a special *Jew oath*, often standing on a pigskin, on the assumption that Jews were essentially perfidious, that is, treacherous and prone to perjury. An additional intent was to insult the Jew

by forcing him to stand on the skin of an animal considered by the Jew an abomination and unclean, and, in the Christian mind related to the Devil. The version of the oath employed in Frankfurt as late as 1847 read similarly, "the Jew shall stand on a *sow's* skin" with his hand on the Torah. The religious nature of this peculiar oath was confirmed by its provision: "may a bleeding and a flowing come forth from you and never cease, as your people wished upon themselves when they condemned God, Jesus Christ, among themselves, and tortured Him and said: 'His blood be upon us and our children.' "[59]

10. The Church itself sponsored many false accusations against Jews to aggravate the belief that they were the *public enemies* of Christendom *par excellence*. Jews were deicides whose continuing evil was of such magnitude that no crime, even collaboration with the Devil, was too great for them to commit. Jews "defile and corrupt [wine, milk, and] meat with the urine of their sons and daughters" so that Christians would suffer a deadly curse.[60] "Christians shall be excommunicated who because of illness, entrust themselves for healing to the care of Jews" (Council of the Province of Beziers)[61] and "when [Jews] remain living among the Christians, they take advantage of every wicked opportunity to kill in secret their Christian hosts"(Pope Innocent III to King Philip-Augustus of France).[62] All Jews were thieves. Jews kidnapped Christian children to sell into slavery. They were forever plotting against Christianity.[63]

11. In England and France, the crime of *bestiality* was attributed to Christians who had sex with or married a Jewish woman, the legal opinion being that "coition with a Jewess is precisely the same as if a man should copulate with a dog."[64]

12. As early perhaps as the thirteenth century, the Christian myth of the *Wandering Jew* (later, the Eternal Jew) began to circulate throughout Europe, although it was not truly popularized until the early seventeenth. The Wandering Jew supposedly had refused help to Christ en route to his crucifixion and who was therefore cursed to wander the world until the end of time as an example of his sin. The Wandering Jew has become the theological symbol of the Jew, his continued suffering serving as a witness to Jesus' sufferings and death and as a warning to unbelievers.[65] As with so many other Christian myths about Jews, the Wandering Jew fantasy also confirmed, in the Christian mind, the validity of the life, suffering, and death of Christ at the same time as it affirmed Christian identity and the pariahship of the Jews.[66]

13. The apocalyptic prophecy was that "Satan will be released when a thousand years have passed." This was interpreted to mean that the long-delayed Second Coming and Final Judgment of Christ was about to take place, preceded by a battle between the forces of Christianity and those of

the Antichrist. Moreover, early in the eleventh century all Christendom was disturbed by increasing fears of an invasion of Moslem infidels, with whom the Jews were sometimes identified.[67]

A long Christian tradition associated the Jews with the Antichrist. Based on passages in the Gospels of Matthew and Mark in which Jesus warned his followers about the arrival of false messiahs and a time of chaos,[68] the Church Fathers believed that the Antichrist would first come to the Jews because they were the ones who had refused to believe in the true Messiah, Jesus Christ.[69] Pope Gregory the Great presented the Jews as a nation of beasts and devils' disciples who preach the Antichrist. Medieval art and drama was replete with examples associating the Antichrist with the Jews.[70]

According to one prophecy, the Antichrist would be born of a Jewish whore and the devil. Fulfilling some of the expectations of the Jewish Messiah, he would rebuild the Temple of Jerusalem and conquer a worldwide empire for the Jews. At this point Jesus Christ would come a second time, defeat the Antichrist and the people of Israel.[71] One of the idea's most extreme supporters, Pseudo-Methodius, regarded the Jews as soldiers of the Antichrist who lived off human flesh.

The twelfth-century German *Play of Antichrist* portrayed the Jews as mistaking the Antichrist as their Messiah and as saving themselves in the end only by recognizing the truth of Catholicism and converting to it.[72]

About this time, several Latin theologians wrote of the Jews' intimate involvement with the Antichrist and wanted them punished for it. Theologians who commented on this topic included Peter of Blois, secretary to Henry II and later to the archbishop of Canterbury; and the Dominicans Albertus Magnus, bishop of Regensburg and papal legate, and his disciple, Thomas Aquinas, the most influential doctor of the Church.[73]

Pierre de Montboisier, called Peter the Venerable, the tenth-century Benedictine abbot of Cluny, the greatest monastery in the West, addressed the Jews directly as progenitors of the Antichrist: "I am talking to you, you Jews . . . monstrous animal . . . brute beast. . . . You hatch basilisks' eggs, which infect you with the mortal poison of ungodliness. . . . They will be so evilly hatched by you as at last to produce Antichrist, the king of all the ungodly."[74]

14. The *sacred calendar* provided many pretexts for Christian attacks on Jews. On holy days the clergy and the faithful stoned Jews "because they stoned Jesus." Symbolic stoning attacks also took place during religious holidays even when Jews were no longer present.[75] At Beziers, a religious center in southwestern France, where the stoning occurred repeatedly at the beginning of Holy Week, one sermon stated to the faithful, "you have around you those who crucified the Messiah . . . . This is the day on which

our Prince has graciously given us permission to avenge this crime. Like your pious ancestors, hurl stones at the Jews."[76]

Jews did attempt to buy their way out of the public slapping of Jews during Easter time, in several locations in France in "revenge for the death of Christ."[77] Jews were also forced to act as executioners or have gallows erected in Jewish cemeteries; they were on occasion denied burial; and publicly slapped and ridiculed. After being sentenced to death, they were sometimes suspended upside down to prolong their agony, or burned between two dogs—a reverse observance of Jesus' crucifixion.[78] In 1466 Pope Paul II forced young Jews to race every day of the Roman Carnival. The Jewish participants, racing in the nude, were ridiculed, harassed, and humiliated. "The Holy Father stood upon a richly ornamented balcony and laughed heartily."[79] As a substitute, the fourteenth-century pope Clement IX allowed the Jewish community of Rome to pay a cash tribute, at a ceremony in which the prostrate chief rabbi of the Jewish community was forced to extend his neck to be stepped on by a Roman magistrate, the Keeper of the Eternal City. This custom was not abolished until 1848.[80] But in Rome and other Italian cities, ghetto rabbis were forced to dress like clowns and march through the streets while Christian crowds laughed and threw stones.

15. In an attempt to establish the orthodoxy of Christian doctrine, early Christian writers associated Jews with *heretics*. Many Latin and Greek Church Fathers saw heresy as essentially Jewish, though most tolerated the Jews in an inferior and degraded status while seeking the total eradication of Gentile heretics.[81] They held that a heretic was a person whose intelligence and will had been perverted by the devil or his Jewish agents.[82] Tertullian explained that "from the Jew the heretic has accepted guidance in this discussion [that Jesus was not the Christ]. Let the heretic now give up borrowing poison from the Jew . . . the asp, as they say, from the adder."[83] Ambrose argued that all Jews were heretics because Jews should have recognized in Jesus their Messiah.[84] Anastasius called Christian heretics "Jews," and they in turn insulted other heretics as "Jews."[85] Several apocryphal Acts and Gospels also identified Jews and Christian heretics. There is some truth in this association of Jews and heretics, for in response to Christian depredations, the Jews may have supported some Christian heretics, such as the Arian Christians, whom orthodox Christians classed with the Jews as the most unrelenting and dangerous enemies of truth.[86] The Theodosian Code defined heretics as "those who shall attempt to do anything that is contrary and adverse to the Catholic sect."[87]

Opposition to the Church by heretics, pagans, and schismatics during the Middle Ages was frequently reduced to Jewish opposition.[88] The English prelate Robert Grosseteste gave the Middle Ages its clearest definition of heresy: "an opinion chosen by human faculties, contrary to Holy

Scripture, openly taught, and pertinaciously defended."[89] That is, heresy consisted of misreading Scripture of one's own free will, and publicly teaching, professing, and stubbornly adhering to this misrepresentation after correction. Although Jews were obviously not Christians and therefore could not *de jure* become heretics, Jews *de facto* fit this definition of heretic.

Although the papacy did not "officially" consider the Jews heretics, its practical policy did not adhere too slavishly to its logic. The net effect of theology and dogma, canon law and Church policy, was that the Church saw Jews as the standard of evil by which its enemies were to be judged—Moslems, prostitutes, witches, bad priests, and heretics.[90] The Church saw to it that these "enemies" should be marked in public, stigmatized, so that they could be kept from Christians, abused, and execrated.[91] Heretics may have been wrong on some theological points, but the Jews were wrong on all.[92] Yet the Church did not "officially" consider the Jews heretics. Had they done so, the Jews of Europe most likely would have been exterminated during the Middle Ages, as other heretics had been. Nevertheless, identifying the Jews as the worst of unbelievers served as a pretext for committing violence against them.[93]

16. The *Franciscans* and *Dominicans* accused Jews of promoting heresy by attracting Christians to their religion (judaizing), associating with relapsed Jewish apostates, and identifying themselves with any and all heterodox ideas, which the mendicants were charged to root out.[94] The friars regarded the Jews as perverse deviants from the truth of Christianity, which they knew but deliberately and maliciously rejected. Based on this perception, the only alternatives the friars saw for the Jews were conversion, expulsion, or death.[95] They may have been the first to acknowledge the idea of a Europe free of Jews.[96] The thirteenth-century Dominican friar Thomas Aquinas, "the Angelic Doctor," who has been acclaimed as the greatest of Catholic thinkers, took the position that the Jews were not only *servi* (slaves) of the Church, but they were also the enemies of Christianity who at the same time supplied its roots. Thomas added the great weight of his scholastic authority to the position that heretics should be excommunicated by the Church and burned by the secular authorities.[97] Aquinas held that the Jews' behavior was no longer determined by the precepts of Scripture but instead by the Talmud, a work written by those "malicious" rabbis who had murdered Jesus.[98]

Paraphrasing the most antisemitic of the Church Fathers, John Chrysostom, Aquinas referred to Jews as "Those who blasphemed against the Son of Man . . . had no excuse, no diminution of their punishment. [T]he Jews were forgiven their sin neither now nor in the hereafter, for they were punished in this world through the Romans, and in the life to come in the pains of Hell."[99] Aquinas thus reintroduced a conception of the Jews that had been prominent but not dominant in pre-medieval Christian thought: the Jew as deliberate unbeliever, one who knew the truth of Christianity, who maliciously refused to accept it, and who should therefore be punished for it.[100] His ideas became the ideological basis of the Christian assault on the Jews as heretics, as the devil's disciples, and as Christianity's greatest spiritual and physical enemies.[101]

One of the most influential preachers of the fifteenth century was the Franciscan Berthold of Regensburg. Attracting huge audiences for more than thirty years in central and Western Europe, Berthold sermonized against the Jews. On occasion, he seemed to adopt the Augustinian approach to the Jews' presence in Christendom as the Witness People, reminding his listeners not to harm them. But his association of Jews with those whom Christendom directly sought to exterminate—heretics and pagans—and his virulent rhetoric had a greater impact. Berthold associated the Talmud with heresy, and the Jews with the Antichrist and the devil as the public enemies of Christendom; and he argued that it was a crime to let the Jews survive. The "Talmud is completely heretical, and it contains such damned heresy that it is bad that [the Jews] live."[102]

The Proto-Crusades Before 1096

It has not been only in recent history that national or local crises bring on attacks against Jews. Violence against Jews seemed to occur after the agonies of natural disasters, as well as during times when Europe experienced religious crisis or calls to avenge the death of Christ. The Jewish situation worsened considerably in the eleventh century.[103] *Peregrinatio* (pilgrimage) and *milites Christi* (soldiers of Christ) were the Latin terms for Crusade and Crusaders used until the mid-thirteenth century. The term *crucesignati* (the Cross-signed, or Crusaders) almost never occurred before the third Crusade at the end of the twelfth century. The English word "Crusade" and the German "*Kreuzzug*" did not exist until the eighteenth century.[104] Well before the formal Crusades began, the anti-Jewish violence of 1007–12 signaled that the relatively good early-medieval relations between European Jews and their Christian neighbors were coming to an end. In the early eleventh century, some Christians believed that Jews, along with the devil, were responsible for urging the Moslems under Sultans Azed and Hakim to sack the Holy Sepulchre in Jerusalem and in a more general way for conspiring to destroy Christian society.[105] The conspiracy theory may have been prompted by the popular Christian belief that participation in Judaism was an act hostile to Christianity. As Paul had observed in

Galatians 5, Gentiles who become followers of the Torah "are severed from Christ; you have fallen away from grace. . . . A little leaven leavens the whole lump."

As a result, many Jews of France and Germany were coerced into conversion, expelled, or killed. The Latin chronicler, Radulphus Glaber, although satisfied that many of Orleans' Jews were killed because the Moslem attack on the Holy Sepulchre was somehow their fault, was thankful that a few Jews escaped, so that they could "provide a constant proof of their crime [and] bear witness to the blood of Christ."[106] Jewish sources attribute the attacks in France to the instability caused by the growth of Christian heresy and by the desire of King Robert to establish a uniform Christian faith within his realm.[107]

In 1063, 33 years before the First Crusade's infamous slaughters of Jews in France and Germany, French "holy warriors" charged into Spain to fight against the Moslems. En route, they stopped off at Narbonne in the south of France near the Spanish border long enough to attack local Jews, and then continued into Spain to make further assaults on the Jews residing there. They were accompanied by Benedictine monks armed with an extensive array of anti-Jewish literature. These and other instances reveal that by the eleventh century anti-Jewish Christian attitudes had grown much more dangerous than before.

Yet in 1063, Pope Alexander II sent letters of praise to those officials who had protected Jews. He distinguished between the treatment of the political enemy, the Moslems, and the religious enemy, the Jews. But his message was equivocal. For at the same time that he praised those officials who protected Jewish lives from the French soldiers, he nevertheless wrote that the Jews "are by God's grace slaves who, having lost their homeland and their liberty, are living in agony over the whole earth, suffering perpetual punishment, and damned due to their spilling of the Savior's blood."[108] The few papal letters of protection issued during this period and later were ignored because they were too late, too ambiguous, or too ambivalent.[109]

The Origins of the First Crusades

The French aristocrat Odo of Largery was elected pope in 1088 and took the name Urban II. Already in contact with most of the forces that had engaged in holy war against Moslems in Western Europe,[110] Urban arrived in France in early August 1095, preaching Crusade in and around a dozen French towns for three and a half months before his celebrated attendance and speech at the Council meeting at Clermont, just north of Paris, in

November. After Clermont, Urban appeared at another two-dozen locations over a nine-month period, before returning to Italy in September 1096.[111] Thus for more than a year, Urban traveled all over France, except for the extreme northwest and northeast, preaching Crusade during the worst period of anti-Jewish persecutions in France and Germany, the first half of 1096.[112]

Although Urban evidently did not directly preach persecution of the Jews, his speeches seemed to sanction the Crusaders' release of pent-up emotions, many of them anti-Jewish.[113] Many Crusaders may have been confused by the traditional papal position that the Jews should suffer degradation and yet not be killed. Besides, neither Urban nor any pope acted to stop the pogroms during the First Crusade. Because of this, the Jews remembered Urban II as the instigator of the deadly First Crusade, "Satan . . . —the pope of wicked Rome—[who] circulated a pronouncement among all the gentiles . . . . [T]hey took evil counsel against the people of God. They said: . . . Their ancestors crucified their god. Why should we let them live?"[114] Also remaining silent on the massacres of Jews was Antipope Clement III (Guibert). Ironically, Clement took the time in 1098 to condemn Emperor Henry IV for permitting forcibly baptized Jews to return to their Judaism.[115] (During the Middle Ages only a handful of Jews voluntarily converted, whereas hundreds of thousands were forcibly converted.)[116] By their silence, these First Crusade era popes (antipope Clement III, antipope Theodoric, antipope Adalbert, antipope Silvester IV; Popes Urban II, Paschal II, Gelasius II, Callixtus II) were signaling that they saw the end of Jewishness, by murder and conversion, as a good thing. Not until 1119–24 did Pope Calixtus II publish a bull of protection, far too late to help the Jews who had been slaughtered more than twenty years before.[117]

At the end of the council's meeting, Urban had urged a huge audience of prelates and knights[118] to put an end to the internecine wars that were plaguing Christian Europe politically, economically, and morally, as well as to help the Christian Churches in the East.[119] Urban chose the Crusaders' warcry, "God wishes it!" (Deus lo volt!) as well as the sign of the cross as their badge.[120] To Urban's expressed motives, most Crusaders added the determination to recapture Jerusalem, Christ's Holy City, from the Turks.[121] The Jewish chronicles and a Latin writer—Solomon bar Simson, Eliezer bar Nathan, Anonymous of Mainz, Obadiah the Preselyte, and Guibert of Nogent—support the notion that after the pope's speech, many Crusaders asked why not attack the Jews, Christ's greatest enemies living among Christians, before setting out to fight the Turks in the Holy Land.[122]

When Urban II granted the Crusaders absolution from penance in his speech at Clermont, he tapped into the most important motive driving Christians into Crusade. He issued a crusading indulgence, "a judicial act of grace," that allowed Crusaders to free themselves from having to perform penance for their sins. Soon, however, an almost universal popular belief emerged that all sins, truly confessed, past, present, or future, would be forgiven simply by taking up the cross. Although it probably went further than Urban intended, the papacy did not refute this Crusader interpretation of spiritual and material rewards.[123] The soldiers of Christ (*milites Christi*) could now take up the cross and find immediate salvation, no matter what they did to the Jews. The chronicle of Mainz Anonymous accused the Crusaders of circulating a report that "Anyone who kills a single Jew will have all his sins absolved."[124] The mass hysteria that followed Urban's speech was soon translated into the slaughter of Jews.

Whereas the popes did nothing to help the Jews, some secular princes did come to their aid. German emperor Henry IV, who had already granted the Jews of Speyer and Worms special charters in 1090, attempted to stop these attacks on Jews. Many German emperors before and after Henry were protective of their Jews, but often in vain.[125]

The First Crusade 1096–99

The First Crusade was composed of three major movements:

1. In the context of ten years of almost unrelieved drought, flood, and famine, thousands of the poor, dispossessed, and desperate (*paupers*) forming the so-called Peoples' Crusade flocked to the spell-binding preacher Peter the Hermit, also called Peter of Amiens and Peter of Archery, a "miracle-working ascetic" (a *propheta*), who claimed to have a Heavenly Letter authorizing him to call a Crusade.[126] These Crusaders formed in France and then Germany in March 1096, made it to Constantinople in July, and shortly afterward most were destroyed by the Turks. But Peter himself survived and joined the knightly Crusaders passing through the city. He ultimately attended the conquest of Jerusalem in 1099, delivering a stirring sermon to the Crusaders on the Mount of Olives a week before the city fell.[127] Peter's French contingents, which may have been responsible for the widespread murder of Jews before they left France, probably came from areas visited twice by the pope, in 1095 and 1096.[128] Although most of his followers were ordinary people, several of his captains were experienced knights, including Walter Sansavoir and Fulcher of Chartres, who was to rule the County of Edessa, one of the Crusader States.[129]

2. Other Crusaders from France, Flanders, and western Germany were also inspired by Peter's preaching. Peter the Hermit's words have not been preserved, but considering the horrified reaction of the Jews of Trier to his sermons, they probably contained a powerful antisemitic message. In 1099

a preacher before the walls of Jerusalem delivered a sermon probably like those Peter preached three years earlier.

> Rouse yourselves, members of Christ's household! Rouse yourselves, knights and footsoldiers . . . . Give heed to Christ, who today is banished from that city and is crucified; and with Joseph of Arimathea take him down from the cross; . . . and forcefully take Christ away from these impious crucifiers. For every time those bad judges, confederates of Herod and Pilate, make sport of and enslave your brothers they crucify Christ. . . . [W]hat is worse, they deride and cast reproaches on Christ and our law and they provoke us with rash speech.[130]

Although the speech was assumably directed against the Turks at the end of the Crusade, in all likelihood Peter and others had preached such sermons in Europe at the start of the Crusade in 1096. Many in the audience would have understood the above sermon as applying not just to the Turks but also to the Jews, especially with the references to Herod and Pilate. The Crusaders believed that the Jews, more than the Turks, continually insulted Christ and Christians, especially because they refused to accept Jesus as their obvious Messiah. Riley-Smith has noted the danger in presenting complex moral ideas to laypeople in simple terms.[131]

Some of these German Crusader armies were led by priests; two, named Gottschalk and Volkmar, attacked Jewish communities in Wessili and Prague.[132] But the most destructive of the German Crusader armies were led by the important south German noble, Count Emicho of Leiningen, who, after slaughtering helpless Jews in the Rhineland, was himself routed in Austria at Wieselberg, a few miles to the west of Vienna.[133] These German Crusaders never got out of Europe, the remnants defeated by other Christians in Eastern Europe.

3. A baronial crusade, called the First Crusade, was the kind envisioned by Urban II. It was led by several major baronial figures such as Hugh of Vermandois, Robert of Normandy, Stephen of Blois, Robert of Flandfers, Bohemond of Taranto, Godfrey of Bouillon and his brothers Eustace and Baldwin, and consecrated by the presence of the papal legate Adhémar, bishop of Le Puy. The two outstanding barons were Raymond de Saint-Gilles and Godfrey de Bouillon. Raymond de Saint-Gilles, count of Toulouse, was the richest Crusader and later recognized by most as the commander of the attack on Jerusalem. His force consisted primarily of Provençals and Burgundians, arguably the largest contingent of Crusaders. Adémar and Raymond were seen as leading the true sons of Israel, the Christian Crusaders, to the promised land; they were referred to by a contemporary Latin chronicler as Moses and Aaron.[134] Godfrey of Bouillon,

idealized by the chronicler Albert of Aix, was one of the first to leave on Crusade; his troops consisted of Lorrainers, northern Frenchmen, and Germans. He became the first Crusader "ruler" of Jerusalem.[135]

The Moslem and the Jewish Enemy

Crusader enmity toward the Jews often surpassed their antagonism toward the more politically threatening Moslems, the other great foe of Christ.[136] The Jews were, after all, the enemy who was geographically closer and politically and militarily weaker than the Moslems. Besides, although Moslems controlled the Holy Land, Jerusalem, and the Holy Sepulchre, the Jews were guilty of a much more grievous injury to Christ than the Moslems, his crucifixion. During the time of the First Crusade, some Crusaders asserted: "Behold we travel to a distant land to war against the kings of that land. We endanger our lives in order to kill or to subjugate all the kingdoms which do not believe in the Crucified. How much more then [should we war against] the Jews who killed and crucified him?"[137] This Crusader comment was made also at Rouen, where men gathering to take up the cross asserted: "We have set out on a long march across vast distances against the enemies of God in the East [Moslems], even though we have right here, before our very eyes, the Jews, the worst enemies of God. To ignore them is preposterous and foolish." Thereupon, the Crusaders drove the Jews into their synagogue and slaughtered them with their swords except perhaps those who accepted Christianity.[138] Some of the children may have been seized from parents who survived the slaughter and forced into Christianity. This evidently did occur at Worms, Monieux, and other Rhineland communities, and later at Jerusalem.[139] These Crusaders then proceeded to join the Crusade led by Robert Duke of Normandy in October 1096.

Fifty years later, Peter the Venerable noted that the Jews were more deadly than the Moslems because they were the surreptitious foe. In another 50 years, Pope Innocent III put the idea as follows. "[The Jews] have threatened us, as they have done to all their hosts, 'like a mouse in our pocket, like a snake in our lap, like a fire in our insides.' "[140]

Crusade, a New Kind of Violence

The First Crusades reflected a new kind of violence toward Jews. Christian attacks on Jews in earlier centuries had been more limited. But in the late eleventh century, the fundamental anti-Jewishness of the Christian theology of glory combined with enthusiastic Christian militarism. As a result, Crusader assaults were territorially widespread, religiously motivated (sometimes mixed with an economic component), savage and murderous, characterized by communal Jewish martyrdom, and an ambivalent attitude on the part of secular and religious authorities.[141] For the first time, attempts were made to eradicate Jews and Judaism from the face of the earth.

The Economic Motive

During the First Crusade (1096–99), although Pope Urban may have mentioned the riches of the East, he did not order any economic protection for Crusaders. Later popes, starting with Eugenius III during the Second Crusade (1145–53), and the Fourth Lateran Council (1215) declared a dozen times that Crusaders did not have to repay interest owed on their debts. Twelfth-century documents do not single out Jewish moneylenders, but thirteenth-century documents do.[142]Only one Latin chronicler, Albert of Aix, emphasized the Crusaders' economic motives in attacking Jews during the First Crusade. He noted that the Crusaders "massacred the wandering Jews, who were the enemies of Christ, much more because of greed for their wealth than because they were subject to God's justice."[143] Hans Mayer has claimed that the Crusaders' self-proclaimed religious justification for killing Jews—that the Jews should be punished as the primal enemies and murderers of Christ—"was merely a feeble attempt to conceal the real motive: greed."[144]

Admittedly, greed, envy, and need were motives for Crusader attacks on the Jews. Because many Crusaders needed supplies and cash, they extorted them from the Jewish communities on their line of march, and Crusader massacres of Jews were often accompanied by looting. The Latin and the Hebrew chronicles bear witness to these events and to the readiness of the bishops and townspeople also to accept Jewish bribes.[145] Mistakenly believing that canon law allowed the looting of the goods of infidels, many Crusaders claimed that the Jewish enemy should pay for their holy pilgrimage.[146]

Although economic considerations may have been a general motivation for the Crusading impulse itself, especially among the Normans, looting the Jews, however, was a minor consideration. The baronial Crusaders and Emicho's army did not seem to be suffering desperate need or hunger. Had the Crusader motivation been essentially economic, they would not have slaughtered thousands of Jews but only taken their goods or extorted money from them. Nor would the Jews have so often killed themselves if the issue had only been money.[147] Neither Christians nor Jews acted or reacted as if the essential Crusader motivation was economic. As the great French historian Henri Peyre has put it, "No great event in history has been due to causes chiefly economic in nature . . . ."[148]

The Religious Motive

During the Crusades beginning at the end of the eleventh century, traditional Christian hostility toward Jews became radicalized, with the Jews coming more and more to represent an alien enemy residing in the heart of Christendom, a word used more and more to express the sense of polarization of Christians against non-Christians.[149] A despised minority of Christendom's "greatest sinners," scattered and unarmed, barely protected by the era's most powerful authorities, the Jews easily fell prey to Christian Crusaders contemptuous of the Jewish spirit. The Jewish communities hardest hit by Crusader attacks—Worms, Mainz, and Cologne—were the greatest Western European centers of Jewish intellectual and cultural vitality.[150]

Riley-Smith and Robert Chazan argue, and they are supported by almost all of the chronicle material of the period or just after, that the Crusaders thoroughly believed that the Jews must be punished as the enemies of Christ.[151] The Jews' very existence tested the Crusaders sense of Christian identity.[152] Albert of Aix, the Crusader chronicler whose work was used by Mayer to support the economic hypothesis, also noted that French, English, Flemish, and Lorrainer Crusaders claimed that their massacre of Jews was "the start of their expedition and of their service against the enemies of the Christian faith."[153] The very oath the Crusaders swore was based on Jesus' command that "If any want to become my followers, let them deny themselves and take up their cross and follow me."[154]

Religious antagonism also manifested itself when the Crusaders attempted to force Jews to convert. Some Crusaders put it bluntly, "either the Jews must convert to our belief, or they will be totally exterminated—they and their children down to the last baby at the breast."[155] The Christian chronicler Ekkehard of Aura—he accompanied Crusaders in 1101—described Emicho as a brute who, once "called to religion," took up the cross and slaughtered Jews.[156] As Langmuir observes, "it was because they were Christians that Emicho and his followers focused their anger on Jews."[157] His 12,000 Crusaders campaigned throughout the cities on the Rhine, Main, and Danube Rivers, "either utterly destroy[ing] the execrable race of the Jews wherever they found them (being even in this matter zealously devoted to the Christian religion) or forc[ing] them into the bosom of the Church."[158] In May 1096, Emicho's Crusader army massacred the Jewish community of Worms in a second assault. Many Jews killed themselves. A small remnant was "converted forcibly and baptized against their will."[159] A few days later, the Crusaders slaughtered hundreds of Mainz' Jews, stripping them naked and throwing them, "still writhing and convulsing in their blood," from windows until the dying Jews were piled in heaps on the ground. At this point, some Crusaders asked the Jews if they wished to be baptized. When they shook their heads "no" and looked to the heavens, "the Crusaders then killed them."[160]

But if the Crusaders had wanted simply to convert Jews, surely Jewish children would have been saved and raised as Christians. This was seldom done. At Monieux, for instance, the Crusaders murdered Jewish adults and kidnapped Jewish children without demanding ransom. The Crusaders later Christianized them.[161] Instead, the Crusaders often seemed intent on destroying all the Jews, rather than baptizing them.[162] The Crusaders may have decided simply to murder all the Jews because the Crusaders correctly reckoned that the Jews would return to their "Jewish vomit" as soon as practicable. During the time of the First Crusade, Sigebert of Gembloux (d. 1112) indicated that those Jews forced to convert soon returned to their Judaism.[163] Besides, the Jews' reputation as stubborn, stiffnecked reprobates led many Crusaders to believe that the Jews could easily ward off the oceans of baptismal water required for a real conversion. With authentic conversion a lost cause, then all the Jewish enemies of Christ might as well be killed.

On occasion, the Crusaders seemed to kill Jews spontaneously, as a result of a rumor or an accident having befallen a Crusader,[164] but more often the Crusaders claimed revenge for the Jews' crucifixion of Jesus—aggravated by the Crusaders' feudal conception of revenge against an offense to their kin or suzerain.[165] At Mainz, the Crusaders asked, "Why should we let them [the Jews] live? Why should they dwell among us? Let our swords begin with their heads. After that we shall go on the way of our pilgrimage."[166] The Crusaders then approached the Jews' houses. "When they saw one of us, they ran after him and pierced him with a spear."[167] At Mainz also, Count Emicho "showed no mercy to the aged, or youths, or maidens, babes or sucklings—not even the sick . . . killing their young men by the sword and disemboweling their pregnant women."[168] Albert of Aix confirms the premeditated nature of Emicho's assault on Mainz, adding that "Emicho and the rest of his band held a council and, after sunrise, attacked the Jews . . . with arrows and lances. . . . They killed the women also, and with their swords pierced tender children of whatever age and sex."[169]

In the first of two assaults on the Jews of Worms, the Crusaders and burghers unsheathed their swords and killed all the Jews they could, saying, "Behold the time has come to avenge him who was crucified, whom their ancestors slew. Now let not a remnant or a residue escape, even an infant or a suckling in the cradle."[170]

The citizens of Cologne "suddenly fell upon a small band of Jews and severely wounded and killed many; they destroyed the houses and synagogues . . . ."[171] In the aftermath of this attack, as the Jews attempted to flee Cologne at night by boat to the town of Neuss, 20 miles north along the

Rhine River, "the Crusaders, discovering them, inflicted upon them a similar slaughter and despoiled all their goods leaving not even one alive."[172]

In other religiously symbolic actions, Crusaders burned synagogues (often with Jews in them) at Monieux, Rouen, and Jerusalem—the Crusaders planned to destroy the Jews at prayer on the Sabbath in the Speyer synagogue but the Jews anticipated the Crusaders, and so the slaughter took place outside the synagogue[173]— and destroyed Jewish cemeteries.[174] At Mainz, Worms, Cologne, and Trier the Crusaders invaded the sanctums of Judaism where they trampled the Torah scrolls into the mud, tore them, and set them afire before the eyes of the weeping Jews.[175] The final attack of three on Trier in June 1096 was perpetrated by Crusaders arriving to celebrate Pentecost, a holiday commemorating "the Holy Ghost's descent upon the Apostles."[176] The celebrated Crusader Tancred charged Judas' 30 pieces of silver per Jew when he needed ransom money.[177]

For essentially religious reasons, the Crusaders appeared intent on destroying Judaism or Jews one way or another. In the spring and summer of 1096 Count Emicho of Leiningen and his German army, soon to be joined by more Germans as well as French, English, Flemish (some of them Tafurs, whom the Moslems called "living devils," who ate human flesh),[178] and Lorrainer Crusaders under the control of important nobles and seasoned leaders waged an unprecedented exterminatory war against Jews in the cities of the Rhineland, "a first Holocaust." The Crusader bands who massacred Jews were not mobs gone wild but were disciplined bands led by experienced soldiers who saw the Jews as religious pariahs whom Christ wanted dead in revenge for their insults to him.[179] There may have been some confusion between Jews and Moslems, but most Crusaders seemed able to make the distinction clearly.

During the first half of 1096, Crusaders attempted, and often succeeded, in hunting out and extinguishing the Jews of the Rhine River cities of Speyer, Worms, Mainz, and Cologne; and later the Jewish communities in Regensburg and Prague. In France as well, Crusaders attacked Jewish communities. Emicho's men assaulted Jews in the Lorraine cities of Metz and Trier on the Moselle; French Crusaders attacked the Jews of Rouen possibly in September 1096; finally, Crusaders under Raymond of Toulouse probably massacred Jews in Monieux in October 1096 en route to the Alps.[180]

Mainz Anonymous records the Crusaders as saying to each other, "We take our souls in our hands in order to kill and subjugate all those kingdoms that do not believe in the Crucified. How much more so [should we kill and subjugate] the Jews, who killed and crucified him."[181] Solomon bar Simson reports the Crusader war cry as: "Behold journey a long way to seek the idolatrous shrine and to take vengeance upon the Muslims. But here are the Jews dwelling among us, whose ancestors killed him and crucified him groundlessly. Let us take vengeance first upon them."[182] In France, Crusaders proclaimed, "it is unjust . . . to allow enemies of Christ to live in their own land."[183] The nobleman Dithmar would avenge "the blood of the crucified one by shedding Jewish blood and completely eradicating any trace of those bearing the name 'Jew.' "[184]

Religious hostility was also obvious when, at Jerusalem in 1099, soldiers of the First Crusade slaughtered Jewish men, women, and children as other Crusaders had done in Europe. The Crusaders at Jerusalem were under the command of Raymond of Toulouse and Godfrey of Bouillon. Before departing on Crusade, Godfrey of Bouillon "swore wickedly that he would not depart on his journey without avenging the blood of the Crucified with the blood of Israel and that he would not leave 'a remnant or residue.' "[185] But while in Europe, he only extorted funds from the Jews of Cologne and Mainz.[186] Nevertheless, he took his revenge on the Jews of Jerusalem by permitting his troops to murder them along with the Moslem residents of the city. As the eminent Anglo-Catholic historian, Lord Acton, described the Crusader actions at Jerusalem, "the men who took the Cross, after receiving communion, heartily devoted the day to the extermination of Jews."[187] Burning some of the Jews alive in the great synagogue, the Crusaders marched around it singing, "Christ, we adore Thee." They then proceeded to the Church of the Holy Sepulchre, singing in joy and exaltation of their victory as the confirmation of Christianity.

> O new day, new day and exultation, new and everlasting gladness. . . . That day, famed through all centuries to come, turned all our sufferings and hardships into joy and exultation; that day, the confirmation of Christianity, the annihilation of paganism, the renewal of our faith![188]

The Involvement of Clergy and Townspeople

During this period, many of the upper clergy, such as the bishops of Trier, Cologne, Mainz, Worms, Speyer, and Prague, made efforts to protect the lives of the Jews. (The bishops of Mainz and Trier, however, used the situation to try to convert the Jews.)[189] When John, bishop of Speyer, heard that the townspeople and Crusaders planned an attack on the Jewish community, he brought the Jews into his palace and protected them with a large force of his own men. He also arrested some townspeople and "cut off their hands." Although some Jews were killed in a second assault, with the cooperation of the German emperor Henry IV, the bishop hid some Jews from Emicho's army.[190] At Mainz, Jews bribed the bishop, who tried to protect them but in vain. Faced with a united force of townspeople and Crusaders, the bishop fled, leaving the Jews to be slaughtered. Reaching the village of Rüdesheim, the bishop offered to save the Jews if they would only convert to Roman Catholicism. "Either believe in our deity or bear the sins of your ancestors."[191]

Despite their attempts at self-defense, the Jews were no match for the combined forces of the townspeople and Crusaders. In May 1096 "the Crusaders and burghers arose first against the saintly ones, the pious of the Almighty at Speyer." Two days later, at Worms, after a rumor held the Jews responsible for attempting to poison the wells to kill the townspeople, the Crusaders and burghers unsheathed their swords and cried out, " 'Behold the time has come to avenge him who was crucified, whom their ancestors slew. Now let not a remnant or a residue escape, even an infant or a suckling in the cradle.' They then came and struck those who had remained in their houses."[192] At Mainz, most townspeople fought alongside the Crusaders.[193] But a minority of burghers tried unsuccessfully to help the Jews. In the end, the Jews were blamed for the conflict between townspeople and Crusaders and slaughtered.[194] At Cologne, the Jews fled to their Gentile neighbors' houses and were given temporary sanctuary.[195] After the Jews of Trier had bribed Peter the Hermit to leave them alone, the townspeople assaulted them:

> Then our evil neighbors, the burghers, came, jealous concerning what had happened in the other communities of Lorraine: they had heard that great misfortunes had been inflicted and decreed upon the Jews. They [the Jews] now took money and bribed the burghers but all this was to no avail on the day of the Lord's fierce anger . . . [196]

The Second Crusade, 1145–49

Religiously inspired attacks against the Jews did not cease with the First Crusade. More attacks occurred against Jews in France and Germany in 1146, at the start of the Second Crusade. The French Cistercian monk Rodolphe, another *propheta* who was believed to perform miracles and who attracted enormous crowds, echoed the First Crusade's anti-Jewish motif when he preached that the Jewish enemies of God must be punished. About 50 years later, the Jewish chronicler Rabbi Ephraim of Bonn described Rodolphe as "the priest of idolatry [who] arose against the nation of God to destroy, slay, and annihilate them just as wicked Haman had attempted to do so."[197] Anti-Jewish riots broke out in Strasbourg, Cologne, Mainz, Worms, Speyer, Würzburg, and in other French and German cities.[198]

The Second Crusaders, like the First Crusaders, believed their sins would be forgiven if they killed Jews.[199] Some Crusaders repeated what was said 50 years earlier:

> We have set out to march a long way to fight the enemies of God in the East, and behold, before our very eyes are his worst foes, the Jews. . . . You [Jews] are the descendants of those who killed and hanged our God. Moreover, [God] himself said: "The day will yet dawn when my children will come and avenge my blood." We are his children and it is our task to carry out his vengeance upon you, for you showed yourselves obstinate and blasphemous towards him. [God] has abandoned you and has turned his radiance upon us and has made us his own.[200]

Rabbi Ephraim of Bonn, substantiated by the Christian chronicler Otto of Freising, reported that Rodolphe preached, "First, avenge the Crucified one upon his enemies [the Jews] living here among you; then go off to war against the Ishmaelites [Turks]."[201] Although the Jews and responsible Christian authorities were better prepared than during the First Crusade, Rodolphe's preaching was followed by massacres in Western Germany and Eastern France (most Jews were saved by Christians)[202] to the Crusader cry of HEP, HEP (Hierosolyma est perdita, Jerusalem Is Lost).[203]

Rodolphe's demagogy was finally terminated by Bernard of Clairvaux, certainly the greatest spiritual figure, and perhaps the greatest historical figure, of the twelfth century.[204] He was the Church's most respected and influential cleric, the leading figure of the Latin Church, its greatest writer and preacher, a reformer of the powerful and prestigious Benedictine order, confidant of Pope Innocent II, and teacher of Pope Eugenius III, and, like the popes, Bernard believed that religion should control every aspect of society. A child during the First Crusade and sensitive to the Jews' precarious situation during the Second Crusade, Bernard spoke out against the murder of Jews in England, France, and Germany. He warned the English people that "the Jews are not to be persecuted, killed, or even put to flight."[205] An adherent of Augustine's precept about the Jews as the Witness People, Bernard traveled to Germany in late 1146 both to preach Crusade and to hush Rodolphe, "It is good that you go off to fight the Ishmaelites [Turks]. But whoever touches a Jew to take his life is like one who had touched the apple of the eye of Jesus; for [Jews] are his flesh and bone. My disciple Rodolphe has spoken in error—for it is said in Psalms, 'Slay them not, lest my people forget.' "[206] This quotation from Psalms 59:11 continues: "bring them down, O Lord . . . consume them in wrath, consume them until they are no more." These are the same passages that Augustine highlighted in his Witness People prescription for the Jews.

Bernard clearly rejected anti-Jewish violence, but he did not attempt to save Jewish lives because of mercy, charity, or human decency. He told the archbishop of Mainz that Rodolphe's murderous preaching against the Jews was the *least* of his three offenses, namely, "unauthorized preaching, contempt for episcopal authority, and incitation to murder."[207] In Bernard's eyes, as in Augustine's, the Jews were not to be murdered because their suffering serves Christians as constant reminders of Jewish sin and of Christian redemption. "The Jews ought not to die in consequence of the immensity of their crimes, but rather to suffer the Diaspora."[208] Bernard recalled to his English audience that Jews must "remind us always of what our Lord suffered." The Jews are "living witnesses" to Christian redemption, that is why they are scattered all over the world and that is why they suffer. Bernard also notes that at the Second Coming of Christ, when the unbaptized Jews will be redeemed, along with Christians, those Jews who have died before this event will remain in hell.[209]

Bernard may not have expressed the anti-Jewish outrage of his friend Peter the Venerable, but like Peter he based his opinion on the Jews on the theology of triumph. At the same time that Bernard pleaded for Christians not to murder Jews, he weakened his protection by his anti-Jewish rhetoric. Employing Jerome's exegesis of the Gospel attack on the Jews, at one point Bernard noted that the Jews were so cruel that they would have mocked Joseph and stoned the mother of Christ had not Joseph hidden the pregnant Virgin. Likewise, Bernard saw the Jews as hard-hearted and regarded the synagogue as a "cruel mother" who had crowned Jesus with thorns.[210] He used the servile condition of the Jews ("no slavery is as demeaning as that of the Jews"),[211] along with their lack of kingdom, priesthood, prophets, and temple, to demonstrate that the Jews were being punished for history's greatest sin, the crucifixion of Christ. For Bernard, the Jews were venomous vipers whose bestial stupidity and blindness caused them to "lay impious hands upon the Lord of Glory."[212]

Bernard also saw the Jews as the very model of evil. Christian heresy and sin were somehow "Jewish." A Christian who neglected Christ's sufferings was "a sharer in the unparalleled sin of the Jews."[213] In his letter to the English people in 1146, Bernard gratuitously and hurtfully inserted the statement that "Christian usurers Jew worse than Jews."[214] In a letter, he commended the Abbot Warren of the Alps for attacking the indiscipline of churlish monks as "destroying those synagogues of Satan"—a phrase from Revelation. In the same manner as John Chrysostom seven centuries before, Bernard condemned the Jews as ever ungrateful to God and as always resisting the holy spirit, calling them the minions of Satan. In sermons, he preached that "The Jews, ever mindful of the hatred wherewith they hate his Father, take this opportunity to vent it on the Son . . . these wicked men . . . " and that "Judaea hates the Light."[215]

Bernard's distaste for Jews extended to the disputed papal election of 1130. In this year, the majority of cardinals elected Cardinal Pietro Pierleoni, who took the name Anacletus II. Reacting to the fact that the cardinal was the great-grandson of a converted Jew, Baruch-Benedict, Bernard became Anacletus' persistent enemy and an advocate of the small minority of Cardinals whose candidate was Innocent II. Writing to the German emperor Lothair III (d. 1137) in 1134, Bernard took the racist position that "it is an insult to Christ that the offspring of a Jew has occupied the chair of Peter."[216]

Taking advantage of the widespread anti-Jewish hysteria of the period, Anacletus was turned out, replaced by the "purely Christian" Innocent II. Reminding Christians that Anacletus descended from the people who had murdered Christ, Innocent's propagandists exploited current anti-Jewish attitudes. Arnulf, archdeacon of Séez, later bishop of Lisieux, expressed a medieval Christian racism when he argued that Anacletus' Jewishness could be seen in his face and that Anacletus' family "had still not been purified from the yeast of Jewish corruption." Writing to his colleague Norbert of Magdeburg, Archbishop Walter of Ravenna requested that the king and the bishops be warned of the Jewish treachery against God and the holy church, so that "the heresy of the perfidious Jews would be extirpated." Four cardinals calculated that Anacletus had been elected at the very moment that Christ was crucified. One of them was Matthew of Albano, a friend of Peter the Venerable. The protests of honorable men such as Peter of Porto, Reimbald of Liège, and Peter of Pisa could not stand against this wave of malicious antisemitic rumor.[217]

In France, Peter the Venerable, the respected abbot of Cluny, accused the Jews of malicious disbelief.[218] He was enraged by contemporary Jews and Judaism. Jewish disbelief may have raised the same kind of momentous doubts in Peter's mind about the validity of his own Christian beliefs as the doubts of Peter Damian and other contemporary Christians.[219] His subordinate, Peter of Poitiers, wrote to him that "You are indeed the only man of our time who has slain the three greatest enemies of our holy Christian faith, I mean the Jews, the heretics, and Saracens, with the sword of the word of God."[220] Many authors, up to the twentieth century, have defended Peter the Venerable's anti-Jewish polemics. As recently as 1967, Giles Constable, the editor of Peter's letters, wrote that Peter's hostility toward the Jews was not based on racism, and therefore he was anti-Jewish *only* in the religious sense, as if this rendered Peter's words less offensive or deadly in their implications.[221] The Catholic Church historian, Edward Synan, however, has noted that Peter's words are identical in meaning with the First Crusaders' calls to violence against Jews.[222]

Unable to tolerate "the slightest rejection of the Christian faith," during the Second Crusade, in 1146, Peter urged the French king Louis VII to punish the Jews.[223] In so doing, he repeated slogans from the First Crusade.[224]

> What does it profit to track down and to persecute enemies of the Christian hope outside our frontiers, if the blaspheming Jews, far worse than Saracens, not at a distance, but right here among us, so freely and presumptuously blaspheme, ridicule, and defile Christ and all the sacraments of the Christian religion without being punished? . . . "Do I not hate those who hate you, O Lord? And do I not loathe those who rise up against you? I hate them with perfect hatred; I count them my enemies."

As the Church Fathers and later antisemites, Peter persistently employs scripture to justify, explain, enhance, or elaborate his anti-Jewishness. This passage comes from Psalm 139, lines 20–22. French scholar Jean-Pierre Torrell has noted that Peter's language strikingly resembles that used by the First Crusaders at Rouen before their massacre of the Jews.[225]

Peter indicated that in his opinion Moslems were morally and theologically superior to the Jews because Moslems agree with Christians on several theological points, such as the belief that Christ was born of a virgin.[226] Elsewhere, Peter warned the Saracens not to make the same mistakes as the Jews, who have denied the whole Gospel and have confused light with darkness, truth with falsehood, demagogue with prophet, and "who have accepted the Antichrist for a Messiah at the end of time."[227] Peter argued in his letter to King Louis that although Christians and Moslems disagreed on the most important issue, namely, on the "death and resurrection of God and the Son of God," Christians should "hate and curse the Jews, who believe absolutely nothing of the Christ and of the Christian faith and who deny, blaspheme, and ridicule the virgin birth itself and all the sacraments of human redemption."[228]

Peter refers to Psalm 139, which contains lines that advocate a deadly policy toward the Jews: "O that you would kill the wicked, O God . . . . Do I not hate those who hate you, O Lord? And do I not loathe those who rise up against you? I hate them with perfect hatred; I count them as my enemies."[229] Referring to the "ungodly" Jews as "synagogues of Satan," Peter recommended the same punishment for the Jews that God had proclaimed for Cain, a fate worse than death.

> As he did with the fratricide Cain, God would not have us kill the Jews completely, not extinguish all of them, but instead to subject them to greater torment and greater contempt, so that they will experience a life worse than death. . . . Since these slavish, miserable, fearful, wailing, and exiled people had spilled the blood of Christ, they richly deserved a punishment that fit their crime.[230]

Because Peter's abbey had borrowed from Jews, giving them as security "precious objects from the sacristy, in particular the gold plate coverings of a crucifix,"[231] Peter speculated "that the Jews had horribly and detestably abused [these] sacred articles in such a way as to shame us and the Christ."[232] Based on this pretext and on his general position regarding Jewish maliciousness, Peter recommended that the "blaspheming" Jews be expropriated in order to fund the (Second) Crusade against the Moslems.[233] In the nineteenth and twentieth centuries, the French antisemite Drumont and the Jesuit journal, *Civiltà Cattolica*, appealed to Peter's authority in advocating Jewish expropriation.[234] Whereas Pope Eugenius III had ordered only interest remitted and had not specified Jewish usurers, King Louis VII may have been influenced by Peter's letter to permit Crusaders to refuse Jewish moneylenders both interest and principal, in effect, expropriating the Jewish community in France.[235]

In his work on the "Malicious Stubbornness of the Jews,"[236] Peter the Venerable asked whether any people did not think that the Jews were "the vilest of slaves. . . . Those who were once the head have now been converted into the tail of all people."[237] Like John Chrysostom denying the Jews their very humanity,[238] Peter argued that Jews are "beasts, an ox or an ass, the most stupid animal in the world . . . who hear but do not understand. . . . It will be clear, not to Christians only, but to the whole world, that you are indeed such an animal . . . a monstrous animal . . . a laughing stock . . . ."[239] Peter indicated that "the divine spirit has not yet been placed into a hard-hearted Jew; without God's spirit a Jew can never be converted to Christ." Peter doubted "whether a Jew can be human" since he refuses to listen to the plain christological meaning of either the Jewish or the Christian Scriptures.[240]

By the late eleventh and early twelfth century, the major authorities had almost uniformly turned against the Jews. With the introduction of public debates, the creation of the mendicant orders, and the increasing power of a hostile papacy, things would grow even worse for the Jews in the thirteenth century. Although Louis VII resisted Peter's anti-Jewish importunings, the murderous implications of Peter's attitude are obvious. Yet Peter's attitude was surely an important indication of the anti-Jewish atmosphere at the time of the Second Crusade. Robert Chazan has argued that Peter was "the key French figure" in the shift in Christian attitudes toward Jews, who were being seen more and more as resisting the "truth" of Christianity because of their "satanic perversity."[241] In France, "many of the most vicious canards were repeated or even initiated by the leading ecclesiastical and secular figures of the period. When the abbot of Clairvaux [St. Bernard], the abbot of Cluny [Peter the Venerable], and the count of Blois [Theobald] lent their prestige to the spate of anti-Jewish slanders, the Jewish community could properly feel profoundly threatened."[242]

In 1147, just a year after Peter's letter to Louis VII, French Crusaders near Troyes, a hundred miles east of Paris, captured and tortured Rabbi Jacob Ben Meir (Rabbenu Tam),[243] grandson of Rashi[244] and perhaps the leading Jewish Talmudist and moral authority of his time. The Crusaders ripped up a copy of the Torah in his face. Associating Christ's Passion and Jewish guilt, they wounded Rabbi Jacob five times, commemorating the five stigmatic wounds that the Jews had allegedly inflicted on Jesus' hands, feet, and side during his crucifixion. The Crusaders told the rabbi: "You are the greatest man in Israel; therefore we are taking vengeance on you because of him who was hanged, and we are going to wound you just as you Jews inflicted five wounds on our God." Rabbi Jacob was rescued from certain death by a passing knight, to whom the rabbi was released only when the knight persuaded the Crusaders that he could convince him to convert to Christianity. If the rabbi's conversion did not take place, then, the knight promised, he would return Rabbeny Tam to the Crusaders.[245] Twenty-four years after this outrage, Rabbenu Tam memorialized in his writing the burning of the Jews of Blois as "a day of mourning and fasting."[246]

The Third Crusade, 1187–92

At the time of the Third Crusade, prompted by the loss of Jerusalem and Acre and dozens of other Crusader towns and fortresses to Saladin,[247] and led by Philip Augustus (1180–1223) of France, Holy Roman emperor Frederick Barbarossa, and Richard the Lionheart of England, Jews were massacred on the Continent and in England.

In 1187 the Jewish community of Mainz, already having endured slaughter during the First Crusade and beset by accusations of ritual murder during the Second Crusade in 1146,[248] was menaced again by Crusaders frustrated by the catastrophes plaguing the Crusader states in the East. According to the Jewish chronicler Rabbi Moses ben Elazar, the Crusaders threatened the Jews with statements that paralleled those of the first two Crusades: "Behold! The day which we have demanded has come—the day on which to kill all the Jews." Luckily for the Jews, on this occasion city authorities sympathetic to the Jews drove off the attackers. In March 1188, German emperor Frederick I (Barbarossa) in Mainz to take up the cross, supported by the German bishops, granted the Mainz Jews an edict of protection, negotiated, bought, and paid for, by the Jews. "Anyone who harms a Jew and causes an injury, his hand shall be cut off. Anyone who kills a Jew shall be killed."[249] The Jews' defensive preparations, which included the establishment of fortified refuges and appeals to the political authorities for protection, coincided with the Church's desire to control this Crusade better than it had the first two Crusades and resulted in saving Jewish lives.[250] The popes, however, did little at this time. In 1180, Pope Alexander III ordered the bishop of Durham, who was later involved in the York massacre, to take precautions against the Jewish "*superstitio*" and "*perfidia.*" Not until 1188 did Clement III issue a protective bull.[251]

In France, disaster after disaster harried the Jews. For 20 years, from the massacre at Blois in 1171 until the years of the Third Crusade, the French Jews were under attack. Ephraim of Bonn reported that in 1192, after a Christian killed a Jew in the city of Bray, the Jewish community convinced the countess of Champagne to try and hang the murderer, "a [vassal] of the king of France." Using this as a pretext, the expansionist French king Philip Augustus crossed the border into Champagne. As a young king, Philip's mind had been filled by rumors of Jewish ritual murder and blood libel. As the chronicle reported,

> He heard tell many times from other youngsters in the palace that Paris' Jews seized a Christian child on Good Friday and tortured him in their subterranean caves to mock Our Lord who had been crucified on this day: they tormented and crucified him and finally hung him to mock the Christian faith. They had done such things many times during his father's reign and as a result had been tried, convicted, and burned. . . . The king inquired about the truth of this charge and whether such behavior continued to this day. He found that it was true.[252]

The preaching of Fulk of Neuilly, "who detested Jews in all ways," also justified the anti-Jewish policies of this French king against the Jews.[253] Burning eighty or more Jews, Philip destroyed the Jewish community of Bray. His biographer, Rigord, describes this action as motivated exclusively by religious considerations, although it was conditioned by political considerations.

Ephraim of Bonn told of another event. In 1196, a Christian servant of Solomon, a rabbi and toll collector for the duke of Austria, took up the cross but stole money from the rabbi, who had him arrested and jailed. The accused's wife publicly complained in church that her Crusader husband had been imprisoned by a Jew. As a result, Crusaders slaughtered Solomon and 15 other Jews. Although the duke ordered the leaders of the riot to be killed, most of the murderers escaped punishment, as the chronicler wrote, "for they were Crusaders."[254]

In the same year, at Speyer the body of a murder victim was discovered, and the Jews were suspected of complicity in the murder. In revenge, some Christian burghers, possibly with the encouragement of the local bishop, desecrated the body of a recently dead Jewish girl. The townspeople hung the Jewish corpse naked in the marketplace with a rat strung in its hair, "as a mockery and humiliation to the Jews." After the father managed to return the girl's body to its grave, a mob attacked the Jews, killing the rabbi and eight others and burning the synagogue and throwing the Torah in the Rhine. Although Emperor Henry VI intervened and forced the Christian townspeople to make restitution to the Jewish community, several Jews committed suicide. A week later, Christians in the neighboring town of Boppard killed another nine Jews, including the rabbi and cantor.[255]

Other pogroms took place in England during preparations for the Third Crusade. With the death of King Henry II, a formidable protector of Jews, an increase in traditional anti-Jewish propaganda,[256] widely believed accusations of ritual murder and blood libel,[257] and the religious excitement surrounding Crusade serving as background, a series of attacks against the Jewish communities of England began in 1189 and continued into 1190.

The troubles began at Richard I's coronation and grew worse with his departure on Crusade. At his coronation, a Jewish deputation bearing gifts was greeted with cries of "Destroy the enemies of Christ!" and pogroms broke out all over England. William of Newburgh (d. 1198) reported that after a celebration of the mass following the coronation, a Christian struck a Jew with his hand.[258] With this example, many Londoners "began to beat the Jews back with contempt." Although several Jews died in this mêlée, the Jewish situation was to deteriorate further. "An agreeable rumor," commented William, "that the king had ordered all the Jews to be exterminated, pervaded the whole of London with incredible celerity. An innumerable mob . . . soon assembled in arms, eager for plunder and for the blood of a people hateful to all men, by the judgment of God." William ended by noting that God had intended that the Jewish blasphemers, "stiffnecked and perverse toward Christians, should be humbled . . . ."[259]

In his *Chronicon*, Richard of Devizes, the Winchester monk who chronicled both the facts and the attitudes of the murderers, who associated the Jews with bloodsucking vermin, vomit, and feces,[260] wrote that "[Christians] began in the city of London to sacrifice the Jews to their father the Devil . . . . Other cities and towns of the country imitated the piety of the Londoners, and with equal devotion sent down their bloodsuckers with blood down to hell."[261] Many of the Christian rioters were arrested but because of popular attitudes hostile to the Jews, only three were hanged. Of the three, two had robbed a Christian whom they claimed was a Jew; the third had burned down a Jewish house that had accidentally set a Christian neighbor's house on fire.[262]

Also victimized by this London fracas was Benedict of York, one of England's richest Jews, who was severely wounded and despairingly accepted Christian baptism by a monk from York. But when summoned before King Richard the next day, Benedict recanted the baptism. When he died a short while later, his corpse could not find burial in either a Jewish or a Christian cemetery.[263]

In February 1190, the anti-Jewish riots spread beyond the London area. At Stamford and York in March 1190, young Crusaders attacked the Jews as enemies of Christ and then used this charge as a justification for extorting "unjust" Jewish possessions to supply their needs while on Crusade. The riots at Stamford took place at the Lent Fair. According to William of Newburgh, at Stamford "a number of youths who had taken the Lord's sign to start for Jerusalem . . . were indignant that the enemies of the cross of Christ who dwelt there should possess so much when they had not enough . . . . Considering, therefore, that they could be doing honor to Christ if they attacked his enemies, whose goods they were longing for, [the Crusaders saw to it that] some of the Jews were slain."[264]

At York, a Christian crowd, stirred up by Crusading propaganda, unrestrained by the government, motivated by religion and greed, attacked the Jewish community on the Sabbath evening before Passover and two days before Palm Sunday.[265] The massacre began with armed Christians murdering the widow and children of the recently deceased Benedict of York and plundering his house.[266] Most of York's Jews retired to the royal constable's castle, where mutual suspicions led to his ordering the Jews out. Most Jews, perhaps 150, committed suicide, *Kiddush ha Shem*, rather than be killed by Christians. As during the First Crusade massacres, in a religious frenzy the mob murdered those Jews who appealed for mercy in return for baptism.[267] William of Newburgh noted that "without any scruple of Christian conscience, they thirsted for [the Jews'] perfidious blood, aroused by desire of plunder."[268] Clergymen, especially impoverished priests, "who had personally held the Jews in hatred" led the mob against the Jews, according to William.[269] Many Jews "were butchered without mercy. . . . Equal zeal inflamed all, for they thought they would be doing a great act of devotion to God."[270] Employing Augustine's doctrine of the Witness People, William explained that Jews, although enemies of Christ and Christianity, should not be slain since it is a matter of "Christian utility" that

> the perfidious Jew, the crucifier of our Lord Christ, is allowed to live among Christians as the form of the Lord's cross is painted in the Church of Christ, for the continual and most helpful remembrance by all of the faithful of our Lord's Passion, and while we curse the impious action in the case of the Jew, . . . the Jews ought to live among Christians for our use, but serve us for their own iniquity.[271]

The government, concerned for law and order, took a 100 hostages from the city but fearful of taking extreme measures in behalf of Jews never put any citizens on trial or punished them.[272] R.B. Dobson believes in the predominance of economic and perhaps political motives in the York massacre, agreeing with William of Newburgh that a conspiracy of "indebted and pitiless landlords" who felt victimized by the government caused the slaughter.[273] But William of Newburgh and Dobson describe a combination of motives, including the background of religious anger and fear, the active participation of clergymen in the riots, and the fact that the Christian mob was "in the grip of religious frenzy."[274]

Thirteenth-Century Crusades

The first half of the thirteenth century has been called the "golden age of crusading," with Crusades occurring nearly every decade. In this century, the papacy often reacted to a threat to the Church by directing a Crusade against it.[275] Jews were often caught in the crossfire. In 1203, Crusaders burned down Constantinople's Jewish quarter.[276] Two hundred French Jews were murdered during the Albigensian Crusade of the early thirteenth century. When Béziers fell to the Catholic forces in July 1209, many Jews were massacred along with Albigensians and, in the frenzied killing, even Roman Catholics. When the Crusading abbot of Citeaux, later archbishop of Narbonne, Arnold Amaury, was asked how to distinguish heretics from Catholics, he replied, "Kill them all! God will know his own."[277]

During the Sixth Crusade and after, several thousand Jews were murdered before the Crusaders left Europe. In 1236, for example, about 3,000 Jews were murdered in Bretagne, Anjou, and Poitou during the preaching of a new Crusade.[278] When Crusades were preached, even when not followed through, Jews were killed: in 1236 in France, England, and Spain; in 1309 in Germany, the Low Countries, and Brabant. As the Benedictine chronicler Gui Lobineau put it: "The Cross [to many] became a burden too heavy to bear."[279]

A cleric and poet captured contemporary attitudes that prompted Crusaders to assault Jews. The Abbé Gautier de Coincy (d. 1236) stated,

> More bestial than naked beasts
> Are all Jews, . . .
> Many hate them, and I hate them too,
> And God hates them as well,
> And all must hate them.[280]

Abbé Gautier further argued that any Christian who tolerated Jews, "Does not deserve to live long." If he were king of France, he would not allow even one Jew to remain in the realm.[281]

In 1233 and 1236, Pope Gregory IX responded to letters from Jews pleading for protection from Christian attacks. He complained to the French bishops about the torture and murder of Jews. In 1233 he noted that to extort funds from the Jews certain lords "tear [the Jews'] finger-nails and extract their teeth, and inflict upon them other kinds of inhuman torments." Some of the French nobles intended "to exterminate the Jews." Because Jews begged Gregory to intervene, because they agreed to live according to canon law and not annoy Christians nor take usury nor "do anything insulting to the Christian Faith," Gregory ordered the bishops to "induce" the nobles not to harm the Jews. Pope Gregory put no teeth into his letter of protection and refused to threaten excommunication.[282] In 1236, Gregory noted the Jews' "tearful and pitiful complaint," and "humbly [seeking] mercy from the Apostolic Throne." He observed that the Crusaders

> try to wipe them [the Jews] almost completely off the face of the earth . . . . They have slaughtered in this mad hostility, two thousand and five hundred of them; old and young, as well as pregnant women. Some were mortally wounded and others trampled like mud under the feet of horses. They burned their books and, for greater shame and disgrace, they exposed the bodies of those thus killed, for food to the birds of heaven, and their flesh to the beasts of the earth. After foully and shamefully treating those who remained alive after this massacre, they carried off their goods and consumed them.[283]

Gregory's "punishment" of the offenders hardly matches the severity of their crimes. Although he condemned the Crusaders' massacre as an "unholy cause," Gregory would only order French bishops to see that the murderers were to compensate the victims families. "After giving due warning you may use ecclesiastical punishment without appeal."[284] As papal historian Shlomo Simonsohn has noted, "it is . . . difficult to see what [satisfaction] the murdered men, women, and children could derive from that."[285] In another letter on the same day to French king Louis IX, Gregory condemns the Crusaders' crimes as "unspeakably and terribly offensive to God" and refers to the Jewish victims as people created in the image of God. But Gregory's suggested punishment is only to "Force the Crusaders to restore to the Jews all that has been stolen."[286]

The events were confirmed by two chronicles, one Christian, the other Jewish. The Latin chronicler Guimar reported that the Crusaders aided by Catholic clergymen forced the duke of Nantes to forgive the Crusaders for their massacre of the Jews and exile the remaining Jews. The Jewish chronicler Ibn Verga: "In [1236] all the wicked ones . . . took counsel against the sons of our people to change their honor. But they held on to their faith, and sanctified the Heavenly Name, and more than three thousand of them were killed . . . ."[287]

The Shepherds' Crusade

In 1315 perhaps the worst famine in European history occurred. The long-lasting suffering and dislocations resulted in a series of riots in France that focused in 1320 into a movement called the Shepherds' Crusade. The "Crusading" shepherds (*pastoureaux* or *pastorelli*) were attracted to communities of Jews, who were regarded as Europe's primary evildoers. Although the Jews had been expelled from France by Philip the Fair, Jewish communities still existed in the English territories in the southwest. Once the attacks had begun, the shepherds found many sympathizers among the Christian townspeople when they did violence to Jews and their property.[288] Although there may have been economic motives for such collaboration, the primary cause was religious. At Albi, the city officials tried to shut the *pastoureaux* out, but the shepherds forced their way into the town shouting that they had come to kill Jews; they were exuberantly greeted by the Christian townspeople. In other towns, officials collaborated with the populace and the Crusaders in their murderous activities, until hardly a Jew was left alive in the area.[289] Those who survived were most likely baptized. The *pastoureaux* continued into Aragon, where they murdered Jews in Jaca and Montclus. James II of Aragon and King Philip V of France, rulers relatively tolerant of Jews and resistant to ecclesiastical pressure, finally sent troops to put the Crusaders down, but not until after 120 Jewish communities had been destroyed. James and Philip were the only rulers to take an active role in stopping these murderous attacks against Jews.[290]

In 1321, during the Shepherds' Crusade, false rumors were put forth that lepers were poisoning wells in order to kill all Christians. The Jews of France and Spain were believed part of the conspiracy and perhaps an additional 5,000 were consequently burned in the south of France.[291] To make matters worse, in the same year Pope John XXII expelled the Jews from the papal territories in the south of France.[292]

Jewish Responses to the Crusades

The much-discussed Jewish mass suicides during the First Crusade, acts the Jews called *Kiddush ha Shem*, or Sanctification of (God's) holy name, had never been normative in Judaism. Assertive resistance—the acceptance of martyrdom short of suicide—had been the norm when Jews were faced with persecution. Exceptional behavior, such as the mass suicide at Masada in 73 C.E., had been deliberately omitted from traditional rabbinic Judaism. Here, the Jewish suicides had been Zealots, a fringe group of Jewish outlaws. But

Ashkenazic Jews were respectable Jews, merchants, scholars, rabbis; men, women, children, and babies at the breast.[293] Although Jewish Scripture supports the supreme value of human life and the rabbinic prohibition of suicide—"You shall live by [the commandments] and not die by them"[294]—another passage—"You shall not profane my holy name"[295]—was used to justify this behavior. This latter passage has been interpreted by Jews as requiring them to opt for death rather than a public[296] renunciation of the faith.[297]

Kiddush ha Shem seemed to occur when Jews felt that the Crusaders had resolved to kill them all without permitting any Jews to live, or when Jews decided that they would rather die than convert or than allow themselves to be desecrated by Crusader weapons. The choice between forced conversion or bodily death was for many Jews a "choiceless choice" that required them to die as Jews one way or another.[298] Jewish chronicles described the *Kiddush ha Shem* that took place amidst the Crusader slaughter of Jews at Mainz under Count Emicho on 27 May 1096.[299]

> The precious children of Zion, the children of Mainz, . . . offered up their children as did Abraham with his son Isaac. They accepted upon themselves the yoke of the fear of heaven, of the King of kings, the Holy One, blessed be he, willingly. . . . They stretched forth their necks for the slaughter and commended their pure souls to their Father in heaven. The saintly and pious women stretched forth their necks one to another, to be sacrificed for the unity of the Name. . . . They sacrificed each other until the blood flowed together. [Likewise] the blood of husbands mingled with that of their wives, the blood of parents with that of their children, the blood of brothers with that of their sisters, the blood of teachers with that of their students, the blood of bridegrooms with that of their brides, the blood of cantors with that of their scribes, the blood of infants and sucklings with that of their mothers. They were killed and slaughtered for the unity of the revered and awesome Name.[300]

But it was only then when the Jews saw that they were being slaughtered, presumably without being offered the escape of baptism, that they began to kill each other, "preferring them to perish thus by their own hands rather than to be killed by the weapons of the uncircumcised."

Ironically, the numerous occasions of suicide among the Jews when faced with slaughter and forced conversion disturbed the Crusaders enormously.[301] What kind of people, they must have wondered, would slay their children and themselves rather than become members of the true faith? Were they not truly of the devil? A Latin Christian chronicler, Bernold, wrote that the Jews preferred death to baptism because of "the devil and their own hard hearts."[302]

Many Jews died fatalistically or emotionally killed themselves with the prayer, the Shema, on their lips. The Jews of Worms and of Mainz, "when the enemy came upon them, they all cried out loudly with one heart and mouth: "Hear O Israel! The Lord is our God; the Lord is one." Affirming Jewishness and bearing public witness to the truth of Judaism, the Shema (*Shema Yisroel, Adonai Elohenu, Adonai Echod*) is the essential Jewish prayer, quoting Deutronomy 6:4–9. It is recited daily and before death.[303]

Some Jews tried to fight back against tremendous odds. At Worms, a young Jewish man, R. Simhah, killed three Christians who tried to baptize him.[304] At Mainz, Jews "donned armor and strapped on weapons[,] then drew near to the gate to do battle with the Crusaders and with the burghers." Also at Mainz, shortly after Crusaders had discovered a Torah scroll and torn it to shreds, and just before the Jews were killed by the Crusaders, Jewish men and women stoned a Crusader to death.[305]

Other Jews responded with anger. In the midst of the Crusader slaughter, the Jews frequently used invective against their Christian murderers.[306] Their language reflected the increasingly powerless and degraded status of medieval Jewry, as well as the increasing violence perpetrated against it. In the Hebrew chronicles written in response to the Crusades, "Satan" was now the "pope of wicked Rome." Jesus was "the Crucified, a trampled corpse," the "offshoot of adultery," and a "crucified bastard," "wretched idol," "loathsome offshoot." As the Crusaders killed the Jews of Worms and Jews were also killing themselves, they cried out: "Look and see, God, what we do for the sanctification of your great Name, rather than to abandon your divinity for a crucified one, a trampled and wretched and abominable offshoot . . . a bastard and a child of menstruation and lust."[307]

At Mainz, the Crusaders stripped the dying Jews, whom they had just slaughtered, of their money and their clothes. Some Jews had barricaded themselves and the Crusaders had stoned them. Jewish women then "cursed and blasphemed the Crusaders in the name of the Crucified, the profane and despised, the son of lust: 'Upon whom do you rely? Upon a trampled corpse!' "[308]

The cross was "an evil sign," "an idolatous sign."[309] Crusading Christians were "arrogant" and "barbaric," "fierce and impetuous."[310] Baptism was called Christian "stench"; churches were "houses of idolatry."[311] David, son of Nathaniel, in the midst of the slaughter of the Mainz Jews, angrily called out to the Christian murders, images like those employed against Jews for centuries: "If you kill me, my soul will reside in paradise, in the light of life. But 'you will descend to the nethermost pit,' 'to everlasting abhorrence.' In hell you shall be judged along with your deity and in boiling excrement, for he is the son of a harlot."[312] Much of this Jewish anti-Christian invective was an attempt to preserve Jewish identity in the face of Christian violence.[313]

Other Jewish chroniclers regarded writing down the events of the Crusades involving Jews as assertive acts of remembrance, recalling archetypal precedents.[314] Sometimes, the Jews blamed themselves. Since God was just, then all this punishment must be the result of Jewish sin. Both Solomon bar Simson and Mainz Anonymous wrote, "God is a righteous judge, and we are to blame."[315]

There were also calls for divine vengeance.

> On a single day . . . one thousand and one hundred holy souls were killed and slaughtered, babes and sucklings who had not sinned or transgressed, the souls of innocent poor people. Wilt Thou restrain Thyself for these things, O Lord? It was for You that innumerable souls were killed! May you avenge the spilt blood of your servants, in our days and before our very eyes—Amen—and speedily![316]

At other times, the Jewish writer simply lamented the momentous sense of loss of these dead Jewish people.

> Gone that day was the crown of Israel,
> gone were the students of the Torah,
> gone were the outstanding scholars.
> Gone was the glory of the Torah.
> . . . Gone were the fearers of sin,
> gone were the virtuous men;
> gone were the radiance of wisdom and purity of abstinence;
> gone was the glory of the priesthood and of the men of perfect faith;
> gone were the repairers of the breach,
> gone were the nullifiers of evil decrees,
> gone were the placaters of the wrath of their Creator;
> gone were many who give charity in secret.
> Gone was truth;
> gone were the explicators of the Word and the Law;
> gone were the people of eminence and the sage—
>
> all of them gone on this day on which so many sorrows befell us, and we could turn neither to the right nor to the left from the fury of the oppressor.[317]

Ritual Murder and Blood-Libel

The Christian fantasies about Jews as devilish Antichrists, and so forth, had existed from the time of the Church Fathers and persisted to the Nazi Third

Reich and beyond.[318] Beginning in the eleventh century, churchmen became aware of heretical beliefs that attacked their religious sense of self. Church officials, sanctioned by Pope Gregory IX's bull, *Vox in Rama*, accused heretics of abominable beliefs and behaviors that were only imagined by the accusers. The Jews, considered "the incarnation of disbelief in their midst," were at the center of this Christian self-doubt.[319]

It was the popular belief that the deicidal Jews were out to murder Christ in every Christian—Ironically, the Jews were forced by Christian authorities in several instances to act as public executioners of Christians—and that any Christian blood was, in essence, Christ's blood.[320] In London in 1244, all the Jews of England were penalized. Medieval Jews were metaphorically defamed as continual slaughterers of Christ, the lamb of God. For from the eleventh century on, the myth of Jewish crime would be transferred from an assassination of the adult Christ to the Christ child to any Christian child, to any Christian. It was at the same time that the popular Christian imagination began to see the Christ child in the Eucharistic wafer instead of the adult Christ. Ritual murder was attributed by Christians to the Jews because they were seen as the enemies of Christendom and intrinsically hating the true faith.[321]

From the twelfth to the twentieth centuries, accusations of ritual murder and blood libel were made against Jews.[322] Imagined to be in league with sorcerers, devils, and heretics,[323] enemies of God, capable of any crime or sin no matter how great, the Jews were accused of annually murdering a Christian child, usually at Passover-Easter time or sometimes at Puriml[324]—especially when these holidays coincided on the sacred calendar with Good Friday and Easter—using the techniques of *shechita*, Jewish ritual slaughter. This accusation most likely stemmed from the myth of the Jews' unending role in Jesus' crucifixion, which was ritually and maliciously repeated every year by the inherently evil Jews on an innocent Christian child. Rappaport suggests that this Christian myth represents a reversal of the Christian habit of kidnapping and forcibly baptizing Jewish chilren.[325] Medieval Christianity created an alternative explanation for the multiplicity of child murders during the Middle Ages and at the same time comforted the faithful by reminding them that they were on the "right" side of the dichotomy between good Christianity and evil Judaism. The triumph of Christianity and the vindication of Jesus' sacrifice were confirmed by the exposure of the Jews' "crimes" and by their punishment, in which the Jewish evildoers were sacrificed "to an offended deity."[326]

The Jews were also defamed with accompanying blood-libel charges, which asserted that the Jews used the blood obtained from their ritual-murder victim for religious and cultural purposes. Although several authorities—King Louis VII, who had tried to stop the attack on the Jews of Blois,

Emperor Frederick II, King James II of Aragon, and a few popes—were publicly skeptical of these charges, the ritual-murder and blood-libel defamations crop up repeatedly from the Middle Ages until modern times.[327] The Endingen *Judenspiel* commemorated the fifteenth-century ritual-murder charge.[328] The Nazis also used the ritual-murder and blood-libel myths in their propaganda campaign against the Jews.[329]

It was widely believed during the Middle Ages that any Christian blood was, in essence, Christ's blood.[330] This blood, considered by Christians to have sacred-magic properties, was assumed to be sought by the Jews. By 1155, the first ritual-murder accusation had been made and spread by the Benedictine monk Thomas of Monmouth in his book *The Life and Miracles of William of Norwich.*[331] The victim, William of Norwich, died in 1144 at a time of a brutal civil war, just before the Second Crusade, and at the scene of widespread massacre of Jews later in the century.[332] For Brother Thomas, the Jews were "the Christian-slaying Jews."[333] Because the murder had taken place during Holy Week[334] and because the manner in which the alleged martyr had been tortured was especially cruel, it was suspected, wrote Thomas, "that it was no Christian but in truth a Jew who had ventured to slaughter an innocent child of this kind with such horrible barbarity."[335] In 1255, the pious English king, Henry III, was the first official to order the execution of Jews for ritual murder.[336]

Similar episodes soon followed on the Continent. The first ritual-murder defamation on the Continent occurred in Pointoise in 1163 but was soon followed by a more notorious case at Blois, France, in 1171.[337] Although political intrigue was involved, priests incited Count Theobald of Blois to burn 32 Jews, 17 of them women, on unsubstantiated charges that they murdered a Christian child, whose body could not be found.[338] The sixteenth-century Jewish writer and physician of Avignon, Joseph Ha-Cohen, in his *Valley of Tears* described the events. After the disappearance of a young boy, a Christian accused the Jews of Blois with drowning the boy but provided no details. Choosing torture and death over conversion to Christianity, the Jews told each other, "Be strong and courageous for our God, for we are his people . . . ; may God's will be done, for there is no other God."[339]

Although it denied the validity of blood-libel, the papacy never specifically rejected the ritual-murder charges. In contrast, King Louis VII had responded to a petition of the Jews of Paris by clearing them of responsibility for the death of a Christian child, "St." Richard, by issuing a decree of protection, and by condemning the ritual-murder accusation at Blois.[340]

Despite the Jewish prohibitions against shedding innocent blood,[341] the consumption of any blood being seen as an abomination, since "blood is life,"[342] the blood-libel charged that the Jews used the blood of a ritually murdered Christian child for various implausible purposes. The fourth-century Church Father Ephrem had written that beneath the whiteness of Israel's unleavened bread "is shame." He implored Christians "not to accept the unleavened bread of this people whose hands are soiled with blood."[343] Christians believed that Jews used Christian blood as an ingredient in matzot, the unleavened Passover breads, which the Jews then crucified and which bled Christ's blood.

Like lepers and heretics, Jews were associated with putrefaction. It was an evil stench associated with the stigmatized Jewish moral qualities, with the Jewish crime of deicide, with Jewish intimacy with the Devil, whose sulfurous stench rubbed off on them, or with Jewishness in general. The consumption of Christian blood, with its "magic properties," was enough to convert the *odor judaicus* (*foetor judaicus*), the Jewish stink, into a sweet Christian smell, it was believed, just as baptism could cure Jews of the "stench of their unbelief."[344] A thirteenth-century Austrian poet claimed that 30 Jews could stink out even the largest city.[345] In 1263, the Dominican Thomas of Cantimpré, a pupil of Albertus Magnus, confirmed that Jews engaged in ritual murder because they needed the blood of a murdered Christian child to heal themselves.[346] Ever since the Jews of Jesus' time had called out to Pilate that "his blood be upon us and on our children," contemporary Jews had to suffer perpetual slavery. "The curse of the parents falls on the children" (*maledictio parentum currat adhuc in filios*), wrote Father Thomas.[347] The blood libel sometimes charged that a child who was allegedly murdered by Jewish ritual (sometimes the child was reported as having been crucified) was bled by the Jews and his blood used in making Matzos, which in turn were crucified and bled Christ's blood. The blood obtained was also allegedly used to cure hemorrhages or hemorrhoids (brought by the Jews onto themselves as punishment for their refusal to save Jesus from Pilate).[348] In addition, the blood was supposedly employed in the Jewish wedding ceremony, or to help Jewish women gain fertility.[349]

After Inquisitor Conrad of Marburg's persecution of heretics accused of rejecting the Catholic dogma concerning the Eucharist and of conducting abominable secret rites in the 1230s, the Jews of Fulda were accused of ritual murder of five Christian boys on Christmas day 1235. In 1236, after establishing a commission to investigate the matter, the German emperor Frederick II rejected this accusation, holding the Jews of Fulda and the rest of Germany "completely absolved of this imputed crime."[350] In the same year, two other German communities raised similar charges against Jews. The Dominicans complained that he favored Jews over Christians.[351] Chaucer's Prioress also tells the story of Jews who murdered a Christian child who sang of Mary, *Alma Redemptoris Mater*, in 1255. About the same time, the thirteenth-century German monk Caesarius of Heisterbach told the story of a Christian child murdered by Jews because they could not endure his pure song about Mary, *Salve Regina*.[352]

Whereas Christendom was surely the center of medieval European existence, the Jews were not peripheral. They were the necessary theological scapegoat, even when they were not actually present. Several times during Holy Week, when the faithful were reminded of the Passion of Christ, Jews were attacked. Jews were forbidden from the church and even from the street when the priest carried the host.[353] It was the clergy that was usually behind the defamations of ritual murder.[354] In April 1288 in Troyes, the Jews were accused of murdering a Christian on Good Friday as a compulsive reenactment of the crucifixion of Jesus; as a result, Inquisitors tried, convicted, and executed 13 of the most pious, learned, and wealthy Jews of northern France.

A self-proclaimed "scourge of the Jews," the Franciscan Bernadino da Feltre wanted the Jews expelled from Christian society. Preaching that the Jews would soon reveal their evil nature, he delivered a series of Lenten sermons in the mid-1470s in Italy that created anti-Jewish hysteria. He predicted that Easter would not pass without Christians fully understanding Jewish evildoing. Perhaps the most famous of the so-called child-martyrs was Simon of Trent. At Passover and Easter time in 1475, the Catholic preacher Bernadino of Feltre had passed through Trent, in Church-owned territory 75 miles northwest of Venice, warning Catholics that Easter would not pass without their knowing something of the Jewish danger to them.[355] Sure enough, a two-year-old boy named Simonino (later called Simon of Trent) disappeared on Holy Thursday in 1475. The boy's death led to a ritual-murder accusation against the local Jews and some visiting German-Jews. This defamation led to the torture, death, and expulsion of the Jews of Trent.[356] Like the alleged murders of other Christian boys by Jews, Simon's murder was assumed to be a reenactment of the crucifixion of Christ.[357] A little over one hundred years later, in 1582, Pope Gregory XIII beatified Simon as the Blessed Simon of Trent. The Trent defamation went far toward popularizing the myth of Jewish involvement in ritual murder and Host desecration.[358]

Da Feltre's agitation seemed to spread beyond Italy's borders, where ritual-murder trials took place within a few years at Regensburg, Ratisbon, Endingen, Ravensburg, and in Spain.[359] In all three of these trials, the charges were based on traditional theological defamations of Jews, not on the facts of the case, and were employed as effective weapons in the medieval Christian war against the Jews.

In 1494 the accusation against the Jews of Trnava, a town 50 miles east of Vienna, of ritual murder enumerated the purposes that the Jews allegedly had in store for the blood of Christian children:

> First, their ancestral traditions indicate that Christian blood is an excellent cure for the wound of circumcision. Second, they use this blood as an aphrodisiac. Third, because both Jewish men and women suffer from menstruation, Christian blood serves as an excellent remedy when drunk. Fourth, an ancient and secret Commandment obliges them to offer annual sacrifices in God's honor with Christian blood . . . .[360]

These false accusations and antisemitic ideas persisted when hardly any Jews were present, as recorded in England in Chaucer's "The Prioress' Tale" and Shakespeare's *The Merchant of Venice*, as well as in the Low Countries.[361]

Jurists at the University of Ingolstadt maintained as late as 1732 that "it was still a general opinion of this time that the Jews need Christian blood."[362] Despite several papal disclaimers, most lay Christians and churchmen still believed these fantastic libels and thus murdered tens of thousands of Jews. Because many of the defamatory charges often occurred *after* the Jews had been massacred, the defamations may have served a psychological purpose: the victimizers may have projected their own hatred, anger, and shame onto their (now dead) Jewish victims. In the minds of the Christians involved in the legal proceedings based on these defamations, the capture, torture, trials, convictions, and sentences of the Jews and the alleged miracles associated with the victims vindicated the triumph of Christianity and Christians over the powers of evil, that is, over Judaism and Jews. It meant that in the end, Christ was avenged.[363]

Defilement of the Sacred Host

Jews were also accused of using Christian blood in Host desecrations, where the Host allegedly bled under Jewish torture. And they were charged with utilizing both menstrual blood and the Host as part of their concoction to poison wells, thereby causing the plague. (The tendency to blame Jews for the poisoning of wells occurred as early as the First Crusade.)

The Christian connection between the Host, deicide, and the Jews existed long before the High Middle Ages. As early as the fourth century, John Chrysostom had included an analysis of the Jews' inherent evil, their crucifixion of Jesus, and the sacrament of communion. "This is how you [Christians] see him [Christ], touch him, eat him. You want to see his clothes; he not only lets you see them, but eat them, touch them, take them inside yourself. . . . Think how your anger rises at the traitors [Jews], the wrongdoers who crucified him, and take care not to be responsible yourself for the body and blood of the Lord. They, they have killed his holy body . . . ."[364] Also tied into this complex mass of dogma and legend, belief and superstition, was the growing cult of the Virgin Mary. She and the infant Jesus were associated as symbols for forgiveness of sin. This led to new reasons for hostility toward Jews in the Middle Ages and beyond. For not only were the Jews contrasted as symbols of carnal sexuality in general, but also they were considered hostile to Mary because they denied specifically that Jesus was miraculously born of a virgin who had been impregnated somehow by God.[365]

In the seventh century, parallel stories circulated that the Jews had maliciously assaulted the sanctums of Christianity by stealing and then defiling a Christian image or relic. These stories were spread in the West by Gregory of Tours' *Concerning the Glory of the Martyrs* and the East by Byzantine literature. After a miraculous event, such as the bleeding of the icon or the Host, then the Jewish crime was discovered and the Jews were either converted or murdered.

As the boy victims of ritual murder came to be identified with the Christ child, so the ritual murders became fused with Host desecrations.[366] The reports of Jewish desecration of the host (the *sacrosanctum Christi corpus*) were theologically generated myths related to this new cult of the infant Jesus. Because the Jews continued to deny Christ, they were accused of hating all Christians. They were, according to Christian belief, always on the alert for the opportunity to recrucify Christ. No deed was beyond the scope of their evil. Medieval Christians imagined the Jews assaulted the stolen consecrated hosts with knives and needles, sometimes with flame. Of course, the divine host may have bled in sorrow, but it was ultimately immutable, as proved by "miracle" after "miracle." For the host refused to be consumed by flame or otherwise destroyed. In legend, the Jews often converted because the miracle causes them to recognize the "validity" of Christianity.

Accusations of profanation of the Host were also related to the new doctrine of transubstantiation. The same council that had ordered the Christian princes of Europe to mark Jews with a stigmatic emblem, the Fourth Lateran of 1215, declared that in the Holy Communion of the mass, the consecrated wine and the sacred Host were miraculously transformed into the blood and body of Christ. This was no symbolic metamorphosis, but an actual transubstantiation. The Council held that Christ's "body and blood are verily contained in the sacrament of the altar under the species of bread and wine, the bread being transubstantiated into the body and the wine into the blood, by Divine Power."[367] This dogma was further enhanced by a red microbacterium, *micrococcus prodigiosus*, known as "the microbe of bleeding Hosts," that grew easily on wafers left in the dark and produced a blood-red color.

With the new doctrine of transubstantiation and the recent popularity of Jesus as an infant, the worshipper became the consumer of the body and blood of the Christ child. The guilt feelings surrounding this action were, according to Haim Maccoby, then displaced onto the deicidal Jews—already considered "the sacred executioners" of Christ—with the additional characteristics of child murder and consumption of the victims blood.[368] So long as this "miracle" comprised a crucial element of Christian faith, then the deicidal role of the Jews could not be forgotten.

In 1264, Pope Urban IV instituted Corpus Christi as an official feast day for the whole Church to commemorate the miracle that "drops of blood" falling from the bread of communion proved Christ's real presence in the wafer.[369] In 1295, Pope Boniface VIII authorized the construction of a chapel to Jesus Christ. He believed that a Jewish desecration of the host five years earlier, in which the Jews had stabbed and then immersed the Host in boiling water, had caused the water "through a divine miracle to be turned into blood."[370] A few years later, the king, Philip IV the Fair, mentioned it in royal legislation listing Jewish offenses against Christianity. This incident was also celebrated in paintings and decorations in Churches across France.[371]

The feast of Corpus Christi stimulated many Christians to fantasize that the Jews, who obviously denied transubstantiation, nevertheless profaned the sacred Host.[372] The profanation myth imagined that, usually near Passover, Jews hired a Christian accomplice, sometimes a priest, to steal the consecrated Host from a church. The fantasy then portrayed the Jews as secretly bringing the Host to the synagogue, where they assaulted and stabbed it in a recapitulation of their crucifixion of Christ. Someone supposedly discovered the bleeding Host, which could not be destroyed, exposing the Jewish "crime." Sometimes the Jews were portrayed as recognizing the miracle of the indestructible and bleeding Host and thus converting to Christianity. Thereupon chapels, shrines, and pilgrimages were established, and songs, drawings, and pamphlets were created, to commemorate the triumph of Christianity over Judaism. Christians also took their revenge, killing thousands of Jews.[373]

In the high-medieval Mass of the Presanctified, the Host was, in the normal course of events, restored to the altar in an elaborate procession. During this ceremony the consecrated Host was protected by keys and seals most likely because of medieval Christianity's "obsessive fear of Jewish desecration of the sacrament."[374] In 1205, Pope Innocent III accepted a rumor about Jews insulting the Host and wrote to the bishops of Sens and Paris stating, "We have heard that the Jews . . . have demonstrated their insolence by insulting the Christian faith with their abominations, such as forcing Christian wetnurses who take the body and blood of Jesus Christ in their Easter communion to pour their milk into the toilet for three days before they give suck to the Jewish children again. Lest the faithful incur the wrath of God, they should not allow the Jews to do such detestable and incredible things against the faith without punishment."[375]

According to Haym Maccoby, many Catholic worshippers felt guilty about consuming the body and blood of the Christ child in the Host and so displaced their guilt onto the Jews—already considered "the sacred executioners" of Christ—with the associated characteristics of child murder and consumption of the victim's blood.[376]

So long as the "miracle" of the mass, transubstantiation, comprised a crucial element of Christian faith, then the "evil" of the Jews could not be forgotten. In 1290, a French Jew was accused of maliciously torturing the consecrated Host, which resisted his efforts to harm it. In 1295, Pope Boniface VIII authorized the erection of a chapel at the site of the miracle and a few years later, King Philip IV the Fair mentioned the affair in royal legislation. This incident was also celebrated in paintings and decorations in Churches across France.[377]

And Jews who were accused of desecrating the Host were often murdered. As early as the second half of the eleventh century, perhaps during the time of the First Crusade, rumors of Jewish Host profanation led to the slaughter of Jews across France. This may have been the first instance in which all Jews were blamed for the actions of a few.[378] The anti-Jewish Dominican, Giordano da Rivalto (1260–1311), wrote that the Jews continued to murder Christ by stealing the Host and, in a repetition of the crucifixion, attacking it as if it were Christ's body. "[The Jews] are evil at heart and hate Christ with evil hatred; . . . they would, were they able, crucify him anew every day. . . . They are hated throughout the world because they are evil toward Christ, whom they curse." Giordano himself claimed that he was present when the boy Jesus miraculously appeared at a Host profanation by the Jews, and he proudly claimed that this miracle stirred Christians to murder 24,000 Jews as punishment for their evil deed.[379]

The worst massacre of Jews who were charged with profanation of the Host occurred at Rottingen, where Jews were accused of pounding a wafer until blood flowed. Under the age-old and persistent influence of Christian anti-Jewishness, the killing spread over much of central Germany and Austria between 1298 and 1303, when the vengeful followers of the German nobleman Rindfleisch traveled beyond the local area in a "divinely ordained" attempt to murder all the Jews. Some local burghers and bishops vainly tried to protect their Jewish neighbors, as did the Holy Roman emperor Albert of Austria (d. 1308). Again, all the Jews were held responsible for the mythical actions of a few local Jews.[380] Perhaps 100,000 Jews, including converts to Christianity, in 146 German communities were massacred during this period.[381]

Covering the general area of the First Crusade massacres, more murders took place in 1336–37 when bands of "Jew killers" (*Judenschächter*) were stirred to bloody action by alleged Host profanations and anti-Jewish propaganda. The noble called King Armleather (Armleder) exhorted the people to avenge the death of Christ by wounding Jews as they had wounded Christ and by spilling Jewish blood as Christ's blood had been shed.[382] The mass murders were already coming to an end when Emperor Louis the Bavarian intervened in 1337. At the same time, at Deggendorf an alleged desecration of the Host with resulting "miracles" and a desire for Jewish property caused the citizens to murder the city's Jews, and a church of the Holy Sepulchre was built to commemorate the act. Again, the slaughter spread throughout southern Germany and Austria.[383]

In the so-called Passau desecration of 1478, Berthold of Regensberg's anti-Jewish preaching influenced the accusations against the local Jews of conspiring in their anti-Christian wickedness with Jews in Prague and Regensburg.[384] The Jews of Passau were executed, baptized, or expelled. The "miracle" of Passau was that the Host was not, could not be, destroyed by Jews. This demonstrated the living presence of Christ in the Host as well as the triumph of Christ over death. On the site of the local synagogue was built the Church of Our Savior.[385]

In Germany, of the 47 examples of bleeding Host stories between 1220 and 1514, 22 were followed by massacres of Jews.[386] In 1354 the Jews of Seville were attacked for allegedly having desecrated the Host. In Sicily in 1474, the Host-desecration massacres were accompanied by the cry, "Long live the Virgin, death to the Jews."[387] The great artist Paolo Uccello also depicted a profanation of the Host. The work, painted on an altar, was executed for the confraternity of the Sacred Sacrament at Urbino. (The painting shows a Jewish banker allegedly purchasing the Host from a Christian woman and desecrating it. The authorities then arrest the Jew and his family and burn them. The final segment shows devils and angels contesting over the soul of the Christian woman who betrayed the Host.)[388] The Church did little or nothing to stop these massacres. Instead, by allowing priests to preach against the Jews, the Church actually encouraged the murders.

Religious hatred of Jews, of course, was not the only motivation for this defamation. Fear, mistrust, anger, envy, self-doubt were all emotions that may have led to such charges. The reports of Jewish profanation of the Host were, however, fundamentally based on Christian theological anti-Jewishness. Not until Christian officials, such as Nicholas of Cusa, otherwise a harsh antisemite, became skeptical about the validity of miracles associated with the Host profanations did false accusations against Jews begin to diminish toward the end of the fifteenth century.[389]

Modern scholarship rejects any factual basis for these accusations of Jewish Host desecration. Yet the charges were so widespread (four accusations in Germany between 1477 and 1514)[390] and the accusers were so convinced of the truth of this myth that the eminent German scholar and priest, Peter Browe, after concluding that the Jews were not guilty of such charges, still questioned whether "it is impossible that all these accusations were simply based on hatred,"[391] an argument employed, *mutatis mutandis*, in 1929 by Julius Streicher, publisher of the popular National-Socialist *Der Stürmer*.[392]

The Black Death

Christians believed that Jews all across Europe were involved in a kind of international conspiracy, along with the devil, against Christendom. As early as 408 in the Christianized laws of the Roman Empire, Jews themselves had been referred to as a plague. "They want to throw the sacraments of the Catholic faith into disorder. Beware lest this plague proceed and spread widely and contagiously."[393] The Twelfth Council of Toledo in 681 also associated the Jews with the plague. The faithful were admonished to "Tear the Jewish pest out by the root!"[394] During the First Crusade, Jews were accused of poisoning wells and were made to pay for this rumor with their lives. Pope Innocent III wrote to King Philip Augustus of France in 1205 that "when [Jews] remain living among the Christians, they take advantage of every wicked opportunity to kill in secret their Christian hosts."[395] In 1246, the Council of the Province of Beziers claimed that Jews practiced black magic and that Jewish physicians and Jews in general were always out to murder all Christians. And Christians who sought the care of Jewish doctors were also punished. "Christians shall be excommunicated who because of illness, entrust themselves for healing to the care of Jews."[396]

In the mid-fourteenth-century, when one-third to one-half of Europe's population died of the Black Plague, massacres of Jews became chronic. In the tenth century Greek Orthodox Christians asked a prelate to stop the plague from ravaging their town. He replied that he would do so on condition that the town's Jews were expelled so that Christians would no longer be "contaminated by their disgusting practices and the pollution of their religion."[397] Although other groups were at first blamed for causing the plague, Jews were inexorably focused on as the malicious initiators.[398] They were condemned by their Christian neighbors and by the so-called flagellants as being in league with the devil to destroy Christendom through the plague;[399] no matter that the Jews were subject to the plague as well as Christians. In reality, the Jews were victimized by a plague within a plague. At Barcelona, Jews were put to the sword after sermons were delivered against them in the churches.[400] In Switzerland and Germany, thousands of Jews were burned, stabbed, and drowned after being accused of poisoning wells and thereby causing the plague. During this period, tens of thousands of Jews in more than three hundred and fifty European communities were tortured and murdered as conspirators in the destruction of Christendom. As Werner Keller has put it, Christians acted "as if they meant to annihilate the whole of Jewry from the face of the earth."[401]

Massacres, expropriations, and expulsions of Jews occurred in Spain, France, Switzerland, the Germanies, Austria, Poland, Belgium, and Hungary.[402] Berthold, bishop of Strasbourg, president of a meeting of the Council of Towns of Alsace during this period, demanded that all Jews be killed off "*Mit Kind und Kegel*," that is, sparing no one, not even "the children nor the infant in the crib"—similar to language used 250 years earlier during the time of the slaughter of Jews in the Rhineland during the First Crusade, and almost 600 years later by Adolf Hitler. Although the burgomeister Conrad of Wintertur and a few others tried to help, the bishop's demands were heeded. On a Sabbath day, which coincided with Valentine's Day (14 February) 1349, most of Strasbourg's Jewish population of more than two thousand were dragged to their cemetery and burned to death, only those who were permitted to convert saved their lives.[403] Only a few submitted to baptism and were saved.[404]

Mixing accusations of economic exploitation and religious defamation, Christians accused the Jews of poisoning the wells, the most serious of the charges against them. As the contemporary Guillaume de Machaut wrote, "The guilty, malicious, and disloyal Jews . . . poison the rivers and fountains." Although lepers, Moslems, pilgrims, and others were also blamed for the plague, the Jews were most often cited. The lepers were regarded as mere instruments of the malicious Jews in spreading plague. At Chillon in September 1348, Jews were put on trial for spreading plague by poisoning wells at Neustadt with a powder ground up from portions of basilisk,[405] the most frightful of the imaginary medieval animals, with whom the Jews were sometimes associated. The Strassbourg chronicler wrote that although they were horribly tortured, the Jews "never confessed that they were guilty of poisoning." (Nevertheless, they may have confessed under the torture.) Even so, "the people said that the Jews should be burned," and they were, 8,000 to 16,000 of them. Those Jews who were willing to convert to Christianity were allowed to live. Jewish children were seized and forcibly baptized as something "pleasing to God."[406] The chronicler Jean des Preis of Liège also wrote that "It was said by most that this pestilence came from the

Jews, and that they had poisoned the wells and springs throughout the world, in order to spread the plague and poison Christendom. . . . Jews were everywhere arrested and put to death and burned wherever the 'Flagellants' came and went . . . ."[407] The contemporary Jewish poet Barukh ben Yehiel imagined the Christians' shouts at the Jews: "Get away, you filth! the [heathens] cried. Look now, and see: The Jews have polluted and poisoned the well and pit in which water is collected."[408]

Important economic, political, and psychological forces obviously aggravated the religious antagonism to Jews during this fourteenth-century crisis. Politically, the Jews were associated with city authorities against the guilds or with the imperial government against the towns; economically, lay Christians were desperately hostile or downright jealous of Jewish usury.[409] As during the Crusades, however, it was the Jews' misfortune to be the available victims *par excellence*: they were already well established as public enemies of Christendom; they were spread across Christendom, weak, divided, and without legal or significant customary rights; and they had no military or political power of their own. The Church and its anti-Jewish theology were primarily responsible for putting the Jews in this vulnerable position, keeping them there, and providing justification for the Christian assaults.

The Mendicant Orders and Conversion of Jews

The spread of the mendicant friars' anti-Jewish propaganda was also a factor in the degradation of the Jews. Members of these orders, originally only Dominicans and Franciscans, must beg or work for a living and were not bound, like monks, to one monastery. The otherworldly and spiritually pure Francis of Assisi, founder of the Franciscan Order, refused to utter the word "Jew" lest he soil himself with its filth.[410]

By the thirteenth century, under the impact of the mendicant orders, especially the Dominicans, and with the collaboration of secular governments in France and Spain, Jews were forced to listen to sermons that attempted to convince them of the reprobate and disgusting nature of their Judaism and of the importance of conversion to Christianity.[411] Should a Jew convert, Christian joy seemed without limit: Pope Nicholas III "thanked God" and heard "choirs of angels." In 1278 he ordered a Jewish convert to be treated like a "prodigal son, with joy and exaltation."[412]

Much worse, from the Jewish point of view, was the Church's sanction of force with regard to the conversion of Jews. A liberal provision of the Theodosian Code allowed Jews who had converted to Christianity for economic advantage or to escape punishment to return without prejudice to Judaism.[413] Up to the time of the Crusades, although baptism into

Christianity was seen as indelible, the Church followed the Theodosian practice and seldom punished a Jew's reversion to Judaism. But in the twelfth century, Gratian had set a harsher tone by choosing for his collection of canon law most of the anti-Jewish provisions of the legislation of the Church Councils of Toledo. Gratian emphasized not the degree of coercion but the sanctity of the conversion process. "It is only proper that [Jews] be compelled to retain the faith they had accepted [Christianity], whether by force or necessity, lest the name of the Lord be blasphemed and the faith they had assumed be considered vile and contemptible."[414] In the end, the theologians, canonists, and popes adopted the position that Jews forced to convert by Crusader sword or popular riot—and Jewish children baptized without the consent of their parents—would not be allowed to return to their Judaism if they merely passively resisted the conversion procedure. Only if Jews obstreperously objected could the conversion process be challenged and the sacrament declared invalid. But this kind of radical resistance to baptism amounted to death one way or another.[415] Most Jews assaulted by the Crusaders simply refused baptism in the first instance and were murdered or killed themselves in a *Kiddush ha Shem*.

The secular princes were also influenced by the Church in this regard. In 1144, about four years after Gratian had published his collection of canon law (consisting of 4,000 patristic texts, conciliar decrees, and papal pronouncements on ecclesiastical matters),[416] King Louis VII, who had been brought up at the abbey of St. Denis, punished Jews who reverted to Judaism. Louis had been under papal interdict (ecclesiastical punishment excluding the faithful from participation in spiritual matters except for communion) for years and had only recently capitulated to Pope Celestine II. Many European rulers, such as the Spanish James I of Aragón and Peter III, compelled Jews to attend conversionary sermons.[417]

Fifty-seven years later, Pope Innocent III's bull on forced baptism became the specific basis of canon law. Innocent wrote to the archbishop of Arles in 1201 that Jews, including children, who received the sacrament of baptism even "by violence, through fear and through torture," through fraud, or to avoid loss, "may be forced to observe the Christian Faith." Innocent validated forced conversions because Jews baptized willingly or not had already partaken of the "divine sacrament," had received "the grace of baptism," had been "anointed with the sacred oil," and had "participated in the body of the Lord." To release a Jew from conversion would be to blaspheme the Lord and allow the Jews to hold the Christian faith as contemptuous and vile. The only exception to baptism's permanent effect was when a Jew "never consented, but utterly objected" to baptism.[418] Without knowing precisely what he had in mind, the kind of Jewish resistance to baptism Innocent was referring to, surpassing violence and torture, was most likely resisting baptism unto death.

Later in the same century, the fate of Jews who insistently resisted conversion was put in the hands of the Inquisition, an institution controlled by the Dominican and Franciscan Orders dedicated to eliminating Jews by conversion or expulsion. On 7 May 1288, Pope Nicholas IV ordered Hugo de Biondes and Peter Arlin, Dominican Inquisitors in France, to proceed against relapsed Jewish converts as heretics, since forced baptism was valid. During an anti-Jewish riot, in fear for their lives, several Jews consented to the baptism of themselves and their children. After the crisis, the Jews reverted to their Judaism. Despite threats of excommunication and more than a year's imprisonment, the Jews refused to return to Christianity. The probable result was that these Jews were burned.[419] In at least one instance, however, local Christians obstructed the Inquisition's persecution of Jewish converts who were observing Judaism.[420]

The conversionary assault included Jewish children as well. Thomas Aquinas' attitude was ambivalent. One the one hand, he stated that Jews "are not to be compelled to the faith under any circumstances to make them believe, since believing is a matter of free will." On the other hand, Thomas Aquinas did provide justification for a policy of forced conversion when he argued that Jews who have already "received" the faith "ought to be compelled to keep it . . . ."[421]

Aquinas expressed equal ambivalence with regard to the baptism of Jewish children. Opposing the baptism of children, he observed that "No one must suffer injury. Jews would suffer injury if their children were baptized against their will, because they would forgo the parental authority over their children who had become believers. Therefore [these children] must not be baptized against their [parents'] will." But in his dialectical method, Aquinas' arguments in favor of child baptism were very strong: Christians had "a higher duty to save a child from eternal damnation"; since Jews were "the slaves of their Christian prince, . . . their children belong to the prince"; since the Christian God gave Jewish children their soul, He had an even greater claim on the child than the Jewish parent, who provided only its body. And Aquinas' rejection of these powerful and attractive ideas supporting forced baptism of Jewish children was very mildly stated.[422] In another section of his work, Aquinas concluded that only those children who have "the use of reason [and] free will [at seven years old?]" can be baptized without the consent of their parents. This implied that older children could be converted without parental approval. The "guilt" of not baptizing a Jewish child into Christianity rested on the Jewish parent, not on Christians.[423]

Harsher still than Aquinas, other Dominicans, such as Guillaume de Rennes (d. 1264), declared that because Jews were the slaves of the princes, their children could be removed "without any injury" to the parents.[424]

The Franciscan theologian and philosopher John Duns Scotus also advocated forced baptism and Jewish expulsion on the grounds that Jews were not needed in Christian society; their presence only contaminated the Christian soul.[425] If a mere handful of Jews survived on some small island, this would fulfill the traditional theological position that a remnant of Jews would be present at the Second Coming. Arguing like John Chrysostom and in opposition to Augustine, he held that if Jews who were forcibly baptized did not become good Christians, this situation was still better than allowing the Jews to continue practicing their Judaism. The chances were good that the children of these forcibly converted Jews would become faithful Christians.

> The Prince . . . ought to take the little ones away from the control of parents who wish to rear them in a way contrary to the worship of God . . . and he ought to steer them to divine worship . . . . What is more, if these [Jewish parents] are forced by threats and terrors to receive baptism, and after it has been received, to live up to it, I would consider this to have been religiously done.[426]

He believed that only a few Jews should remain as Jews situated on some island to await the Second Coming. Yosef Yerushalmi called this a "medieval 'Madagascar Plan,' " a Nazi plan to ghettoize all European Jews in anticipation of their gradual eradication.[427]

The Clergy's Involvement in Anti-Jewish Violence

Many medieval churchmen, several of whom were canonized, advocated violence against the Jews. Throughout the Middle Ages, the rhetoric of Christian preachers inspired by anti-Jewish theology had a hypnotic effect on their audiences.[428] In tenth-century anti-Jewish riots in France, a tonsured cleric apparently called for mass murder of Jews. He was reported to have stated to the local count and his men,

> Be not like your ancestors who did not learn the proper lessons, and take vengeance against these people, unfortunate and scattered throughout the earth. Thus these people became a great snare for us . . . . This is what you must do to them—desolation and decimation. You must utterly destroy them from the land, leaving them neither root nor branch. Let us send letters throughout all the distant towns, and let them do likewise to this recalcitrant people . . . ."[429]

In his chronicle of the Christian attack on Jewish communities in England in 1190, William of Newburgh described the hysterical anti-Jewish actions of a white-robed Premonstratensian priest "who . . . was busily occupied with the besiegers, standing in his white garment and frequently repeating with a loud voice that Christ's enemies ought to be crushed, and encouraging the warriors by the example of his help."[430]

In the fifteenth century, the Church, and especially the mendicants, continued to inspire antisemitism. One of the Church's great preachers, the Franciscan Bernadino da Siena, was violently opposed to the Jews. He felt that Jewish moneylenders deprived Christians of their wealth and Jewish physicians deprived Christians of their "health and life." He also believed that the Christian doctrine of love with regard to the Jews was only theoretical. "There can be no concrete love towards them."

Another Franciscan, Giovanni di Capistrano (d. 1456), believed like John Duns Scotus that Jews were not really needed in Christendom. Known as "the scourge of the Jews," he was appointed by Pope Martin V as Inquisitor for Germany, the Slavic countries, and Italy.[431] He preached to the faithful and carried on an anti-Jewish propaganda campaign in royal and papal courts beginning in 1417. Instrumental in coercing the Jews to wear the badge of shame and boasting about abolishing "the devilish privileges" of the Jews, he also managed to obtain the cancellation of pro-Jewish charters, or *privilegia* (concessions granted the Jews by local political authorities). These charters were the best protection the Jews had at the time and were probably purchased at great expense by them. It cost the English Jews 4,000 marks to have their charters confirmed in 1201. These *privilegia* were awarded the Jews because the sovereign authority who "owned" them meant to make a profit by taxing Jewish wealth. In 1442, Capistrano convinced Pope Eugenius IV to deal a series of devastating blows to an already beleaguered Jewish community in Spain. He attacked the Jewish means of living by forbidding Jewish moneylending in Spain and Italy; he further alienated Jews from their neighbors by reinforcing the social separation between Christian and Jew; he diminished their safety by abolishing earlier charters in their favor.[432] An intelligent man of many talents, Capistrano debased himself by spreading defamations of ritual murder and Host desecration that led to the destruction of the Jews of Breslau and Passau in 1453. He argued that without baptism the conflict between Christians and Jews could only be solved by sending the Jews, the "enemies of the faith," to their death on the high seas.[433]

Bernadino da Feltre, mentioned earlier, was a disciple of Capistrano's. Despite the generally friendly relations between Christians and Jews in Italy, in almost every Italian city during this period, reforming priests such as Jacobo della Marca, Alberto da Sarteano, Roberto da Lecce, and Girolamo Savonarola (1452–98)—the Florentine Dominican preacher noted for his political activities—preached violence of one sort or another against the Jews.[434]

Spain During the Inquisition

Antisemitism had been prevalent in early medieval Spain when Arianism gave way to Catholicism, and a second wave of anti-Jewish clerical activity covered the thirteenth through the fifteenth centuries.[435] But the collaboration of Church and state between the anti-Jewish riots of 1391 and the expulsion of the Jews in 1492 was unparalleled in medieval Jewish history and not to be witnessed again until the twentieth century.[436]

Although Fritz Baer points out some examples of good relations between Christians and Jews in fifteenth-century Spain—some Christians visited their Jewish friends on Jewish holy days; some Jews served as physicians, stewards, lawyers, state officials, and tax farmers (a position that gained nothing but Christian hostility)—the persistent assaults of the Church and of Christian authorities against Judaism, Jews, and *conversos* were much more characteristic of the era. Jews were forced to live in ghettos, to wear the badge of shame, deprived of synagogues and cemeteries, denied economic rights, defamed by the ritual-murder libel, and attacked by monks who tried to convert them or who led the populace against them.

This persecution was foreshadowed by the preaching of a mendicant friar of Estella, near Pamplona, in 1328, which had led to the deaths of 5,000 Jews. In the same year, in an act of *Kiddush ha Shem*, Jews of Navarre chose to burn their synagogues and homes and drown their children in wells rather than convert or allow themselves to be defiled by Christian swords.[437]

However, these relatively minor persecutions were merely a prelude to the massive anti-Jewish campaign at the end of the fourteenth century led by Ferrant Martinez, deputy for the bishop of Seville and Confessor to Queen Mother Leonora. Like Ambrose, Martinez called for the razing of synagogues "in which the enemies of God and the Church practice their idolatry."[438] When he became administrator of the diocese of Seville in 1390, Martinez again demanded the destruction of Jews and Judaism. His sermons instigated an anti-Jewish movement that spread to most of Spain. The violence, aggravated by the royal authorities half-hearted requests that Martinez desist, began on Ash Wednesday in 1391 and lasted on and off for decades. Christians, both municipal officials and common people, looted and massacred perhaps 50,000 Jews, forcing thousands of others to be baptized.[439] This is how the Jewish chronicler Hasdai Crescas described it:

> the Lord bent His bow like an enemy against the community of Seville . . . [Christians] set fire to its gates and killed many of its [Jewish] people; but most changed their religion, and some of the women and children were sold to the Moslems . . . and many died to sanctify His Name, and many violated the holy covenant.[440]

The Jewish apostate Solomon Halevi, who as Pablo de Santa Maria became bishop of Burgos in 1406, stated that the riots were caused by a "mob frenzied by the blood of the Messiah."[441]

A government inquiry found that not only the common people but also nobles and priests were involved. When the queen requested the bishop of Osma to arrange safe passage for her Jewish subject, Samuel Bienveniste, the bishop noted that he was sure that her majesty would not be displeased should the Jew be converted to Christianity while he was the bishop's guest. Both the Church and Christian princes forced the Jews to remain in Spain during the riots so that the *conversos*, Spanish Jews most of whom had been coerced into Christian baptism, would not be able to revert to Judaism in another location. The *conversos*, many of whom practiced Judaism in secret, were to be isolated from their fellow Jews, who supported them.[442] Spanish Christians considered both Jews and former Jews potential traitors, and the Inquisition treated *conversos* much more harshly than Moslem converts (*moriscos*).[443]

Although Spanish Franciscan Vincent Ferrer proclaimed that "Christians must not kill Jews with knives, they must kill them with words," his passionate anti-Jewish sermons led to mob attacks between 1411 and 1413.[444] He burst into synagogues with his flagellants and converted the Toledo synagogue into a church. Described in chronicles as a "scourge of the Jews," he oversaw 20,000 forced baptisms and pressured John, king of Castile, to expel the Jews unless they converted. His anti-Jewish program, called the Statutes of Valladolid, included ghettos, Jews as pariahs, and forced conversions, which the government of Castile carried out.[445] Ferrer may also have influenced the Avignonese anti-pope Benedict XIII to attempt a mass conversion of Jews by condemning the Talmud.[446]

Exploiting the intimidating anti-Jewish riots of 1391 and their aftermath to convert the Spanish Jews, Benedict called a public debate at Tortosa (1413–14). The pope demanded that the Jews adhere to the Church's regulations isolating Jews from Christians, prohibiting Jews from holding public office, from exacting usury, from building synagogues, and from possessing the Talmud.[447] The Jewish representatives at Tortosa were detained for months by Benedict and their communities were intimidated and physically attacked by preachers such as Vincent Ferrer and by Christian mobs.[448] In mid-1413, Benedict ordered the Jewish representatives to answer charges of heresy and defend the Talmud against charges of blasphemy.[449] As a result, 3,000 Jews converted to Christianity. In the next few years, Benedict also issued several bulls that rewarded Jews for converting and penalized the rest for resisting conversion.[450] Benedict's most extensive bull decreed a variety of restrictive anti-Jewish measures and ordered that any Talmuds found in Jewish hands would cause the Inquisition and ordinary Church officials to proceed against the Jews as heretics.[451]

By mid-century, "Old Christians" were assaulting tens of thousands of "new Christians" (*conversos*). At Toledo in 1449, *conversos* were tortured and burned for their "irreligion." In 1460, a Franciscan monk, Alfonso de Espina called for the establishment of an Inquisition to root out *converso* heresy and Jewish blasphemy.

Isabella and Ferdinand, the Catholic rulers who originally created the Inquisition in Spain to support the Church and to purify Spain of Jewish *conversos* reinforced the Inquisition, which attacked Jews, heretics, Moslems, free thinkers (*alumbrados*), and Protestants as unchristian elements. (As great a Catholic as Ignatius Loyola, who founded the Jesuit Order, was imprisoned twice by the Inquisition.) Combining nationalistic motives with antisemitism to Christianize, expropriate, murder, or expel the Jews from Spain, Ferdinand and Isabella adopted the Dominican Alonso de Hojeda's proposal that only an Inquisition could destroy the *converso*-Jewish influence in Spain. Although the crown accrued enormous economic and political benefits from *converso* and Jewish activities—in Aragón in the year 1294, for instance, the Jews provided the crown 22 percent of all direct taxes, besides special subsidies and forced loans—the Spanish monarchs put religion first.[452] In 1478, the Spanish monarchs petitioned Pope Sixtus IV to clamp down on *conversos* who were only superficially converted, many through coercion. These *conversos*, in the pope's words, "observe Jewish superstition and the dogmas and precepts of perfidy . . . the depravity of heresy. . . . They not only persist in their blindness, but they also infect with their perfidy [their children and spouses. *Conversos* were] a pernicious sect" to be extirpated by the roots. Sixtus then abrogated all privileges the *conversos* had been granted.[453] In 1483, the crown, along with Pope Sixtus, selected the Dominican prior Thomas de Torquemada as Inquisitor General.[454]

Rather than accept new Christians into Spanish life and culture, the Inquisition set out to destroy the Jewish presence. More than 90 percent of the Inquisition's victims were Jews or *conversos* accused of reverting to Judaism and therefore considered heretics. At its most extreme, the Inquisition considered anyone who believed in the Talmud and the future arrival of the Messiah, that is, *any* Jew or Judaizer, a complete heretic who deserved to be burned.[455] The priest-historian Andrés Bernáldez gloated that in the 1480s alone, hundreds of *conversos* were burned by the Inquisition and thousands authentically Catholicized. "The Inquisition proposed to destroy both the belief and the believers. 'The fire has been kindled, and it will burn until not one of them is left alive.' "[456] The Inquisition handed over so-called heretics to the state, the "secular arm," with the hypocritical recommendation that heretics should be executed "without bloodshed," that is, by burning.

Two polemical booklets, written by intimates of Torquemada in 1488, made clear the radical attitude of the Inquisition. They held that in contrast to Old Christians, the New Christians were deceitful traitors who aided the Moslems and were lazy and valueless enslavers of Christians. They would be destroyed along with the Antichrist. Citing Isidore of Seville, one of the pamphlets, *Tratado del Alborayque*, claimed that "in Spain heresy will arise from among those who crucified Jesus. That heresy . . . will be destroyed by fire and sword."[457]

In the spring of 1492, a year after the Inquisition's fabrication of the La Guardia ritual murder—the alleged child victim was never even named[458]—an edict of expulsion against the Spanish Jews, not the *conversos*, was drafted by the Inquisition and signed "by order of the King and Queen, our sovereigns, and of the Reverend Prior of Santa Cruz [Torquemada], Inquisitor General in all the kingdoms and dominions of Their Majesties." The order—issued less than three months after the Catholic monarchs accepted the surrender of the final Moslem stronghold in Spain, making the Iberian peninsula fully Christian for the first time since 711—allowed the Jews a few months to set their affairs in order and leave. It called the Jews representatives of a "damnable religion . . . undermining and debasing Our holy Catholic faith." During the remaining months, church and state made a last-ditch effort to convert Jews: rabbis were detained so that priests could campaign among the Jews, new laws were enacted that benefited converts, promising temporary protection from the Inquisition, and Jews were forbidden to take any significant wealth out of Spain. When a few Christian towns promised to protect Jewish cemeteries from profanation for a fee, the authorities nullified these agreements. Their rationale was that the Inquisition had categorized Jewish property as the same as that of heretics.[459]

Thus the Church and Christian state, and its militant faithful, conquered and destroyed Jews, Jewishness, and Judaism in Spain. Although Isabella and Ferdinand were not above the tactics of self-serving political considerations, *Realpolitik*, the theological doctrines of Christianity profoundly influenced their motives. (King Ferdinand and Queen Isabella were reportedly on the verge of revoking the edict of expulsion because of the Jewish promise of a huge ransom. At this point, Torquemada, who had been the queen's confessor when she was still *Infanta*, entered the court carrying a crucifix, stating "Judas Iscariot betrayed Christ for thirty pieces of silver, and now you would abandon him for thirty thousand. Here he is, take and sell him."[460]) As in France before the 1306 expulsion of Jews, the Inquisition indoctrinated both rulers and populace with Jew-hatred.[461] Like

French king Philip IV, the Spanish rulers contravened their own economic and political interest for the sake of religious conformity. On 19 December 1496, Pope Alexander VI conferred the title "Catholic" on the Spanish king and queen for their unification of Spain and opposition to the French in Italy, as well as their expulsion of the Jews.[462] In 1597, Spanish king Philip II on his deathbed ordered the last 500 Jews to be expelled from Spanish Milan on the grounds that whatever practical use the Jews might be, reasons of faith had to prevail.[463]

The forced exile of the Spanish Jews in 1492[464] and the persecution of the *conversos* revealed the fragility of the Jewish presence in Christendom. The Spanish-Jewish community had been the largest, most prosperous, learned, and powerful in Europe. But now, without the support and comfort of the wider Jewish community, the *conversos* would become totally assimilated into the Catholic population.

Faithful Catholics applauded this expulsion. The priest Johannes Eck justified the expulsions of Jews from France and Spain as understandable considering that the Jews were so evil.[465] Even moderate Catholics and humanists of the time, many of whom knew Jews personally, had no sympathy for the exiled Jews. Instead, often using anti-Jewish theological language, they regarded the exile as a final Catholic judgment on the Jews. Despite his friendship with the Jew Elias del Medigo, the Italian humanist Pico della Mirandola believed that the expulsion of the Spanish Jews represented a victory for Christianity, an exaltation of the Catholic religion through the "Most Christian King [Ferdinand] who is beyond all praise."[466] As secular a figure as Francesco Guicciardini regarded the Spanish assault on Jews as totally justified.[467] In 1512–13, Guicciardini wrote that, under the earlier reign of Henry IV, Castille had been beset by "an ugly and outrageous infection" of Jews and heretics, whose "depravity" corrupted the whole kingdom. If reforms had not been undertaken, then "within a few years all of Spain would have left the Catholic faith." Guicciardini went on to praise Ferdinand and Isabella and the papal Inquisition for solving Spain's difficulties, through their destruction of Jewish culture in Spain. He admired the Inquisition's burning of 120 Jews at Cordova in one day. Should fear of the Inquisition cease, he observed, then the *conversos* would simply "return to their vomit."[468] The seventeenth-century *philosophe*, Pierre Bayle, noted that "to this day [Jews] hate Christians without restraint, considering it good in God's eyes to fool Christians by going to Mass with great exhibitions of zeal in order to avoid the Court of the Inquisition . . . ." Although the Jews admittedly had received "horrible treatment" in Spain, he wrote, "much of it was their own fault; for why did they live there under the guise of Christians horribly profaning all the sacraments, when they could have gone elsewhere proudly to profess their Judaism?"[469]

The Middle Ages established that the most terrible things could be done to Jews, including mass murder, and these actions would be justified by Church doctrine. Economic and political factors undoubtedly aggravated Christian violence against Jews. But Christian antisemitism, continually repeated by theologians, clergy, and artists, was the fundamental cause and justification of the degradation and killing of Jews. As Richard Rubenstein has noted, "Once the fantasy of murdering Jews has become a fact, it invites repetition."[470]

Chapter 4

The Germanies from Luther to Hitler

We are at fault in not slaying the Jews.

—*Dr. Martin Luther*

Reformation Attitudes Toward Jews

The Reformation terminated the Roman Catholic Church's religious monopoly over Christendom, but it did not provide religious freedom for, or improve the status of, the Jews. Almost everyone—peasants, burghers, aristocrats, priests, theologians, and humanists—considered the Jews deicides, enemies of the Churches, a pestilence on society, and murderous thieves intent on stealing Christian wealth and lives. In an era when most Jews lived in miserable poverty, Christians charged that Jews exploited them and threatened to destroy the essence of Christian life. People scapegoated Jews for the harsh fiscal policies of Church and state, the authorities for social protest; most blamed the Jews for the plague and the imminent appearance of the Antichrist, an ally of the Jews.

Great efforts were made to convert Jews in preparation for the anticipated battle between Antichrist and Christ. The German humanist and jurist Ulrich Zasius admonished the princes and all Christians to baptize Jewish children and to force adults to baptism. Reminiscent of John Chrysostom, Zasius further observed that Jews were "truly professed enemies [and] truculent beasts" who should be eliminated from Christian society.[1]

Both Catholics and Protestants detested Jews. Catholic antagonisms followed along the familiar lines of medieval antisemitism, and both humanists and Protestants felt that before Church and society could be fundamentally reformed, they had to be totally cleansed of their Jewish spirit.[2] Protestants blamed Jews for what was wrong with Catholicism, and Catholics blamed Jews for what was false in Protestantism. The Jews served as a scapegoat for the defects Protestants discerned in the Roman Catholic Church, just as the Church saw the Jews behind Luther. The Talmud was attacked on the grounds that it contained blasphemous and seditious material.[3]

Traditional religious antisemitism prevailed among German Catholics.[4] In 1477 the Catholic burgher Peter Schwartz summarized many of these anti-Jewish attitudes when he observed,

> The Jews have been punished severely from time to time. But they do not suffer innocently. They are in agony because of their wickedness, because they cheat us and ruin whole nations by their usury and secret murders. . . . There is no people more evil, more cunning, more greedy, more impudent, more annoying, more venomous, more wrathful, more deceitful, and more shameful than the Jews.[5]

The Erfurt Augustinian Magister, Johannes von Palt, accused the Jews of knowingly and willingly torturing Christ. The vicar of the Augustinian Order and Martin Luther's early superior, Johannes von Staupitz, charged the Jews with inflicting the "harshest" torments on Christ, "savagely" striking him: "O you evil Jew! Pilate teaches you that your character is harsher than a pig's; the pig at least knows mercy."[6]

Luther's Catholic opponent, Johann Eck, a student of Zasius, regarded Jews as murderous by nature: bloodthirsty magicians who reveal the demonic in their behavior. The very image of the Devil, the Jews seek Christian blood, which they need to wash away their stigma for having murdered Christ. "It is no wonder that the Jews now buy the blood of innocent children, just as their fathers had bought the innocent blood of Jesus Christ from Judas with thirty pennies." Eck repeated myths that Jewish men menstruate, that only Christian blood could cure Jewish ills, and that the Jews of Genoa had murdered a Christian child, using his blood as a sauce into which they dipped pieces of fruit to eat. He concluded that Christian persecutions of Jews were justified as self-defense.[7]

Antisemitism was also prevalent among the Protestants. John Calvin, who had rejected the use of force to convert Jews in his early writing, believed that they were collectively guilty of deicide. Calling them "profane dogs,"[8] Jews in Calvin's mind were the worst of people: ignorant, greedy, impious, ungrateful, rebellious, and criminal. He felt that "the Jews stupidly devour all the riches of the earth with their unrestrained cupidity."[9] Like other Christian theologians, Calvin considered the Jews as model evildoers. In studying their behavior and fate, Christians could understand the penalties for disobeying God. Michael Servetus (d. 1553) accused the priggish Calvin of emulating aspects of the Jewish Law, which Servetus called "irrational, impossible, tyrannical. . . . You [Calvin] place the Christians on a par with the vulgar Jews."[10]

Calvin condemned Servetus to the stake and the whole Jewish people to hell.[11] His attitude is perhaps best summarized by this statement: "The degenerate and unlimited stubbornness [of the Jews] has served to justify their unending accumulation of misery without limit and without measure. Everyone seems cheered by their punishment; no one feels sorry for them."[12]

Although no longer bound to the Church, Francesco Guicciardini (d. 1540), the historian and statesman, was, like other humanists, conditioned by traditional Christian antagonisms toward Jews. He applauded the expulsion of the Jews from Spain in 1492. Prizing the efficacy of centralized authority, he judged the forced exile as one of the Spanish monarchy's major accomplishments. He was profoundly disturbed that Jews had occasionally been chosen to administer Castile's finances. He felt that the Jewish presence would have led to the destruction of Spanish Catholicism. As a result, he praised the Inquisition, even when it burned 120 Jews in Cordova in one day.[13]

In France at the same time, the relationship between Abraham ben Mordecai Farissol (d. 1528) and François Tissard, further demonstrated the deteriorating Jewish situation within Christendom. Tissard was a humanist, philosopher, lawyer, classical scholar, and student of Hebrew and Judaism, which he learned from Farissol, a scribe, cantor, community leader, polemicist, Biblical exegete, geographer, and teacher. Tissard wrote extensively of his warm feelings toward Farissol, yet he perceived Jews through the lens of traditional religious hostility.

Tissard wanted to learn about Judaism so that he could better defend the Christian faith and convert the Jews to Christianity. But his attempts to convert Jews through loving kindness ("It's not right that we attack and subjugate them with harshness, warfare, or persecution . . .") still displayed the traditional lack of respect for Judaism. Beneath Tissard's desire to gain Jewish conversion through love lay Tissard's dislike, even disgust, for Jewishness. Although he occasionally commended the Jews of Ferrara—he did respect the Jews' care for their poor—his words were often uncomplimentary and sometimes laced with contempt and animosity. He regarded Jews as hypocritical, lazy, vulgar, and stubborn. Despite his close friendship with Farissol, he believed myths about Jewish men: that they menstruated and their thumbs shook with usurious guilt. He described the Jewish religious service as revolting: "one might hear one man howling, another braying and another bellowing; such a cacophony of discordant sounds do they make! Weighing this with the rest of their rites, I was almost brought to nausea." He perceived the Jews as

> [F]oolish, senseless, insane, and frenzied people! What demon, what raging madness seized you so long ago? . . . Be converted! Be converted, I say! to him . . . who for you as for the rest of mankind atoned for the first sin with suffering and death. . . . Be converted to him, who, on the cross, prayed and pleaded for you, . . . and offers you pardon. . . . But if you oppose him he will destroy you utterly, will rightly, for your hateful thanklessness, disown you to deprive you of your inheritance, and will cast you away, into eternal fire.[14]

Other Swiss and German reformers, such as Ulrich Zwingli and Martin Bucer, although appalled at Luther's abusive vulgarity, still shared his belief that Jews threatened Christian society. Bucer convinced Landgrave Philipp of Hesse to institute discriminatory regulations against the Jews.[15] Only Justus Jonas and Andreas Osiander, who had sharply criticized Luther's anti-Jewish writings, advocated genuine toleration. Osiander (anonymously) defended Jews against the ritual-murder defamation: he cautioned magistrates investigating the murders to ask themselves "whether the priests or monks were not themselves eager to obtain the appearance of greater sanctity, more miracles, and to establish new pilgrimages, or whether they were much inclined to exterminate Jews." Johannes Eck called Osiander an "evangelical scoundrel" who dared to defend the "bloodthirsty Jews." According to Eck, because Osiander denied the historical validity of Jewish ritual murder, he was in effect accusing Christians of murder and lies.[16]

Perhaps the most surprising antisemite of the time was the former priest and eminent humanist, Erasmus of Rotterdam. When he wrote to the Dominican theologian von Hochstraten that "If being an authentic Christian means hating Jews, then we are all good Christians,"[17] he apparently did not exclude himself. Like Luther, his criticism of the Roman Catholic Church never implied that Judaism should be tolerated, and he held that no place existed in Christian society for "the most pernicious plague and bitterest foe of the teachings of Jesus Christ."[18] Erasmus believed that the Jews' stubborn refusal to convert jeopardized the most fundamental values of Christian society.

The Jews in Luther's Worldview

Martin Luther was the most important historical figure of the sixteenth century. The founder of Protestantism was a child of his times and with regard to the Jews a true son of the Catholic Church. Despite devoting only a small proportion of his work to the "Jewish Question," Luther was concerned with this issue throughout his life.[19] Much of his work focused on salvation through faith in Jesus as the Messiah, and the Jews were central to the issue of Jesus' Messiahship. In Luther's view the Jews still rejected Christ through ridicule, still killed Christ through ritual murder, and still awaited their "false" Messiah. Luther saw Jews as "the quintessential *other*," and he placed them far outside his "sacral community."[20] He wrote: "O God, beloved father and creator, have pity on me who, in self-defense, must speak so scandalously . . . against your wicked enemies, the devils and the Jews. You know I do so because of my zealous faith, and in your honor; for the question involves all my heart and all my life."[21]

Luther's enemies' list changed from time to time but usually included his Protestant and Catholic rivals, unruly German peasants, the pope, the Antichrist, the devil, and the Turks, whom he called "Red Jews."[22] Luther sometimes wrote as if the Jews headed the list,[23] although they were among the most vulnerable and impoverished of his contemporaries. The mere fact of their existence, Luther held, presented a threat to Christians everywhere. Jews were active opponents of Christ, and Christians had replaced them as God's people. Jews were a continuing model for mankind's opposition to God[24] and did not deserve toleration.

In one of his earliest references to Jews, a letter of 1514 to another reformer Spalatin (Georg Burkhardt), a thirty-year-old Luther wrote: "I have come to the conclusion that the Jews will always curse and blaspheme God and his King Christ, as all the prophets have predicted."[25] Although Luther became the sixteenth-century's most vehement anti-Jewish writer, "That Jesus Christ Was Born a Jew,"[26] his first full-length discussion of the Jews and Judaism represented only modest opposition to Judaism and Jews. In this work, Luther criticized the Catholic Church's anti-Jewish tradition and harsh measures, although he would later adopt the same antisemitic ideas and actions. He attacked the Catholic Church's blood-libel myth as "foolishness"; he admitted the closeness of the Jews to Jesus and the special relationship they had with God the Father; he advocated an assertive and less aggressive approach to *Judenmission* (the attempt to convert the Jews); and in a self-deprecatory way, he criticized all Christians for not being kind to the Jews.[27]

Despite its appeals for tolerance and understanding, "That Jesus Christ Was Born a Jew" was the opening barrage in Luther's theological assault on the Jews, which was to grow much harsher as time passed. For even though Luther rejected coercion and advocated persuasion to convert the Jews—Hans Sachs called him the "Wittenberg Nightingale"—his stated intent was that the Jews leave their false faith, Judaism, for the "true Biblical faith," for "their own true faith," that is, for Christianity. Luther never advocated respect for Jews as Jews. Instead, he wanted them to stop hoping and waiting for the false Messiah, the *Shiloh*.[28] He may have pursued his *Judenmission* for the greater glory of Christ, for the greater satisfaction of Martin Luther, and for his vision of the good for the Jews themselves. But when these "stiff-necked people" would not be baptized, Luther's honeyed words turned to gall. Twenty years later, he refused to heed his own warnings against anti-Jewish defamations and harsh treatment of Jews, and championed the very things that he claimed to abhor. "That Jesus Christ Was Born a Jew" ended on what proved to be a portentous note: "Here I will let the matter rest for the present, until I see what I have accomplished."[29]

Even before his patently anti-Jewish works of 1543—"On the Jews and Their Lies" and "Vom Schem Hamphoras," reprinted five times during his lifetime[30]—and his last sermon in his hometown of Eisleben in 1546, there were hints that Luther was gradually becoming enraptured by a hatred of Jews. In 1536 or 1537, Josel of Rosheim, appointed *shtadlan*, or "supreme leader and regent of the Jews" by Charles V, delivered a letter to Luther from the moderate Strasbourg Protestant leader Wolfgang Capito. Josel's mission, ultimately successful, was to convince Elector John Frederick not to expel the Jews of Saxony. Josel tried to enlist the aid of Luther, whom Josel blamed for the elector's threats against the Jews, but Luther refused to meet with him.[31] Josel wrote in his memoirs that "[The Jewish predicament in Saxony] was due to that priest whose name was Martin Luther—may his body and soul be bound up in hell!—who wrote and issued many heretical books in which he said that whoever would help the Jews was doomed to perdition."[32]

Luther's Recapitulation of Catholic Antisemitism

Learned, vituperative, obscene, Luther's last major writings on the Jews are an excellent compendium of nearly all the anti-Jewish ideas written, spoken, painted, or sung in the Catholic tradition. Luther claimed that he wrote his ferocious tracts against the Jews because of their continuing attempts "to lure to themselves even us, that is, the Christians."[33] He argued that the Gentiles were the newly and truly chosen people of God; the Jews belonged to Hell. The Jews "surely are no longer God's people. . . . And if they are not God's people, then they are the devil's people." Why? Because they are malicious murderers of the prophets and of Christ himself. "They have scourged, crucified, spat upon, blasphemed, and cursed God in his word, . . . the true Messiah, one whom his own people had crucified, condemned, cursed, and persecuted without end." Their evil was so great that they "would crucify ten more Messiahs and kill God himself if this were possible, together with all angels and all creatures, even at the risk of incurring thereby the penalty of a thousand hells instead of one." Like many Catholic theologians, Luther specifically disparaged Judaism's crucial religious values: "[Jews are] base, whoring people, that is, no people of God, and their boast of lineage, circumcision, and law must be accounted as filth."[34] And this continues for a hundred thousand words.

In more than a dozen different contexts, Luther associated the Jews with the devil. Jews were "the devil's people," "consigned by the wrath of God to the devil," "circumcised physically to the devil," "serpents and children of the devil." Luther recommended that "Wherever you see a genuine Jew, you may with a good conscience cross yourself and bluntly say: 'There goes a devil incarnate.' " (Fifty-two years later, Shakespeare used a similar phrase in describing Shylock: "Certainly, the Jew is the very devil incarnation.")[35] Citing Gospel, Luther wrote: "We will believe that our Lord Jesus Christ is truthful when he declares of the Jews who did not accept but crucified him, 'You are a brood of vipers and children of the devil.' "[36] In *Von Schem Hamphoras* he added that the God of the Jews is the devil. They "have a God who is a master at mockery and is called the accursed devil and the evil spirit."

Luther noted that the Jews denied the ritual-murder accusation but maintained that he believed it true.

> I have read and heard many stories about the Jews which agree with this judgment of Christ, namely, how they have poisoned, made assassinations, kidnapped children. . . . I have heard that one Jew sent another Jew, and this by means of a Christian, a pot of blood, together with a barrel of wine, in which when drunk empty, a dead Jew was found. There are many other similar stories. For their kidnapping of children they have often been burned at the stake or banished . . . . I am well aware that they deny all of this. However, it all coincides with the judgment of Christ which declares that they are venomous, bitter, vindictive, tricky serpents, assassins, and children of the devil, who sting and work harm stealthily wherever they cannot do it openly.[37]

Luther argued that the Jews were even more malicious than the Catholics had thought. As minions not of God but of satan, they had been commanded to murder all Christians.

> I firmly believe that they say and practice far worse things secretly than the histories and others record about them, meanwhile relying on their denials and on their many deceits. . . . They have been thirsty bloodhounds and murderers of all Christendom for more than fourteen hundred years in their

> intentions, and would undoubtedly prefer to be such with their deeds. . . . Their pious conduct . . . is so thickly, thickly, heavily, heavily coated with the blood of the Messiah and his Christians.[38]

Luther's Obscene Anti-Jewish Rhetoric

Luther often expressed his antagonism to Jews in filthy language. He lived when and where obscenity and foul language were frequently used techniques of verbal assault, especially against the Jews. A folk tale described how Til Eulenspiegel, the roguish peasant boy, tricked some Jews of Frankfurt into buying Til's own feces as a medicine.[39] The Jewish Messiah was represented in this dialogue from a German play of 1500: "The biggest mother swine must be brought . . . the [Jewish] Messiah must lie under the tail! What falls down from her he should . . . in one gulp swallow it down." (Feces were widely associated with demonic pollution.) Just a half-century after Luther's death, a Frankfurt broadsheet showed the synagogue littered with emblems of the Jews' "depravity," namely, the Golden Calf, the ritually murdered Simon of Trent, Jewish holy books, and Jews with ugly names and criminal professions. The captions on the broadsheet labeled the Jews:

> The learned fool . . . instructs the point-heads,
> How they should shit on [bescheissen] the Christians.
> . . . the gallows-worthy thief,
> sticks the tube right up the arse.
> . . . . Judas betrayed Christ swiftly,
> And also the child in Trent. . . .
> May the devil ride you . . .
> And in the end the fire [of Hell] burns.[40]

In the midst of this den of Jewish criminality and blasphemy lay the *Judensau*, the "Jew swine." The Jews supposedly learned their Talmud by kissing the *Judensau*, sucking its milk, and eating its feces. The *Judensau*, according to Isaiah Shachar, was to the Christian artist "the symbol *par excellence* of perverted Judaism and Jews.[41]

The obscene talk of street and field pervaded Luther's anti-Jewish rhetoric. The Jews, the Jewish Scriptures, and the Talmud all fell victim to his obscene assault. He called the synagogue "a defiled bride, yes, an incorrigible whore and an evil slut with whom God ever had to wrangle, scuffle, and fight." The popular belief that devils ate feces supported Luther's identification of Jews and devils. "It serves them right that . . . instead of the beautiful face of the divine word, they have to look into the devil's black, dark, lying behind, and worship his stench." Luther also compared the Jewish Scriptures to pig feces—the Nazis made the same analogy[42]—which he alleged that the Jews consumed. "You damned Jews . . . You are not worthy of looking at the outside of the Bible, much less of reading it. You should read only the bible that is found under the sow's tail, and eat and drink the letters that drop from there."[43]

Luther's attack on the Jews grew more foul and obscene in *Vom Schem Hamphoras* written later in 1543. Here, he was the first to describe in writing the *Judensau* carved on the church at Wittenberg, and he further identified the Jews with Judas, whose feces and urine the Jews were also pictured as consuming. More lewd was his claim that the Jews themselves were full of "the devil's feces . . . which they wallow in like swine."[44] Luther also contributed a fecal dimension to Jerome's association of all Jews with Judas.

> When Judas Iscariot hanged himself, his guts burst and emptied. Perhaps the Jews sent their servants with plates of silver and pots of gold to gather up Judas' piss with the other treasures, and then they ate and drank his excreta, and thereby acquire eyes so piercing that they discover in the Scriptures meanings that neither Matthew nor Isaiah himself found there, not to mention the rest of us cursed *goyim* [Heb. for "nations"].[45]

Luther directed the same obscenities to Christians who treated Jews humanely. He wrote in "On the Jews and Their Lies,"

> Now let me commend these Jews sincerely to whoever feels the desire to shelter and feed them, to honor them, to be fleeced, robbed, plundered, defamed, vilified by them, and to suffer every evil at their hands—these venomous serpents and the devil's children, who are the most vehement enemies of Christ our Lord and of us all. And if that is not enough, let him stuff them into his mouth, or crawl into their behind and worship this holy object. . . . Then he will be a perfect Christian, filled with works of mercy—for which Christ will reward him on the day of judgment, together with the Jews—in the eternal fire of hell.[46]

Luther's Program for the Jews

Without real hope that the Jews would be converted, Luther implored the German princes to follow a cruel policy, actually carried out 400 years later by a modern German "prince," Adolf Hitler. Both Luther and Hitler advocated the destruction of Jewish religious culture, the abrogation of legal protection, expropriation, forced labor, and expulsion of the politically defenseless Jews.[47] Luther also urged mass murder of Jews.

Although Luther begins the following passages from "On the Jews and Their Lies" with the observation that "the harsh mercy" of this pogrom must be exercised with "fear and trembling," his language remains extremely violent.

> First, . . . set fire to their synagogues or schools and . . . bury and cover with dirt whatever will not burn, so that no man will ever again see a stone or cinder of them. This is to be done in honor of our Lord and of Christendom, so that God might see that we are Christians . . . . [It would be good if someone could also throw in some] hellfire.
>
> Second, I advise that their houses also be razed and destroyed.
>
> Third, I advise that all their prayer books and Talmudic writings, in which such idolatry, lies, cursing, and blasphemy are to be taught, be taken from them. . . . also the entire Bible. . . . [The Jews] be forbidden on pain of death to praise God, to give thanks, to pray, and to teach publicly among us and in our country. . . . [T]hey be forbidden to utter the name of God within our hearing. . . . We must not consider the mouth of the Jews as worthy of uttering the name of God within our hearing. He who hears this name from a Jew must inform the authorities, or else throw sow dung at him when he sees him and chase him away. And may no one be merciful and kind in this regard.
>
> Fourth, I advise that their rabbis be forbidden to teach henceforth on pain of loss of life and limb. . . .
>
> Fifth, I advise that safe-conduct on the highways be abolished completely for the Jews. . . .
>
> Sixth, I advise that usury be prohibited to them and that all case and treasure . . . be taken from them and put aside for safekeeping. . . . Whenever a Jew is sincerely converted,[48] he should [receive a cash bonus].
>
> Seventh, I recommend putting a flail, an ax, a hoe, a spade, a distaff, or a spindle into the hands of young, strong Jews and Jewesses and letting them earn their bread in the sweat of their brow . . . . For it is not fitting that they should let us accursed Goyim toil in the sweat of our faces while they, the holy people, idle away their time behind the stove, feasting and farting . . . .[49] [The fiction that Jews did not work for a living was found in several Christian authors. Opposed to Jewish usury, Thomas Aquinas felt that "the Jews should be forced to work [the land] and should not be allowed to live only by means of usury as they do in Italy, where they spend the whole day doing nothing." In fourteenth-century Spain, Gonzalo Martinez de Oviedo wrote to Alfonso XI of Castile that the Jews should be expelled for religious reasons, adding "My lord the king goes out to make war against his foes while they sit at home eating and drinking."][50]
>
> The country and the roads are open for them to proceed to their land whenever they wish. If they did so, we would be glad to present gifts to them

> on the occasion; it would be good riddance. For they are a heavy burden, a plague, a pestilence, a sheer misfortune for our country.[51] [This last phrase was employed likewise by the eminent nineteenth-century German historian Heinrich Treitschke, the Pan-German League, and the Nazis.]

Extermination of the Jews?

At the end of World War II, on the basis of the sometimes unreliable evidence of Luther's *Table Talk*, Peter Wiener attacked Luther for his championing of mass murder of Jews.[52] Gordon Rupp defended Luther against Wiener, but Rupp surprisingly ignored the evidence of Luther's major work on the Jews, "On the Jews and Their Lies." Luther himself was not a consistent advocate of killing Jews, but he seemed to want to murder them if all else failed. And he often believed that all else would fail.

Several elements in Luther's program contained murderous implications. Setting fire to the synagogues, homes, and holy books of the Jews, the first three steps of Luther's program might have met some physical resistance from the Jews, which, in turn, would have led to the Christian princes' use of lethal force. Even Jewish resignation to the inevitability of these actions would probably have led to riots and other potentially murderous actions against Jews. Luther also advocated a crude and merciless attack on Jews who pray, which could also have resulted in Jewish deaths.

Luther's program for the Jews asked the princes three times to kill Jews who resisted. His third and fourth steps mention "pain of death" and "pain of loss of life and limb." His fifth step advises the authorities to deprive the Jews of safe passage once they have left their ghettos. Another passage of "On the Jews and Their Lies" indicates that Luther saw the necessity of killing at least some of the Jews:

> I wish and I ask that our rulers who have Jewish subjects exercise a sharp mercy toward these wretched people . . . . They must act like a good physician who, when gangrene has set in, proceeds without mercy to cut, saw, and burn flesh, veins, bone, and marrow. Such a procedure must also be followed in this instance. Burn down their synagogues, forbid all that I enumerated earlier, force them to work, and deal harshly with them, as Moses did in the wilderness, slaying three thousand lest the whole people perish. [They are a] people possessed . . . .[53]

Luther repeatedly employed the language of violence against most, if not all, of his enemies.[54] Since Luther often wrapped his opponents in his vision of the Antichrist, he urged that brimstone and sword be used against many individuals and groups: the pope and his cardinals, Catholic nobles, anti-Lutheran sectarians such as the Sabbatarians, heretics, Turks, and German peasants. Luther's famous recipe for dealing with the peasants, his former allies, went as follows: "Let everyone who can, smite, slay, and stab, secretly or openly, remembering that nothing can be more poisonous, hurtful, or devilish than a rebel. It is just as when one must kill a mad dog; if you do not strike him, he will strike you, and the whole land with you."[55] Whereas the peasants presented a real and physical danger to Luther and the Lutheran establishment, the Jews posed no such physical threat.

Luther did not, however, distinguish between the threat of peasants and Jews because he regarded the powerless Jews as a menace to his theological vision of life, earthly and eternal. Jews became a symbol for all that was evil and devilish, unChristian and anti-Christian in the world. For these reasons, the Jews appeared more threatening than Luther's other enemies, and he assaulted them with his words and advocated governmental violence against them. In at least three other passages Luther approached an advocacy of mass murder of Jews.

A sermon of 1539 argued, "I cannot convert the Jews. Our Lord Jesus Christ did not succeed in doing it. But I can stop up their mouths so that they will have to lie upon the ground."[56] The language is ambiguous, but it implies a death threat. The imprecise language allowed people of goodwill to believe that outright murder was not being proposed, while at the same time this kind of language permitted them to "speak about the unspeakable," the mass murder of Jews.[57]

In "On the Jews and Their Lies," Luther wrote that Jews, like usurers and thieves, should be executed:

> [Today's Jews] are nothing but thieves and robbers who daily eat no morsel and wear no thread of clothing which they have not stolen and pilfered from us by means of their accursed usury. Thus they live from day to day, together with wife and child, by theft and robbery, as arch-thieves and robbers, in the most impenitent security. . . . If I had power over the Jews, as our princes and cities have, I would *deal severely* with their lying mouth. . . . For a usurer is an arch-thief and a robber who should rightly be *hanged* on the gallows seven times higher than other thieves.

The deadly syllogism Luther concocts in this paragraph may be stated as follows:

> All thieves and usurers should be executed by hanging.
> All Jews are thieves and usurers.
> Therefore, all Jews should be killed.

In another section of "On the Jews and Their Lies," Luther clearly stated that all Jews should be murdered.

> We are even at fault in not avenging all this innocent blood of our Lord and of the Christians which they shed for three hundred years after the destruction of Jerusalem, and the blood of the children they have shed since then (which still shines forth from their eyes and their skin). We are at fault in not slaying them.[58]

The Aftermath of Luther's Teaching

Within Germany, Martin Luther's work has had the authority of Scripture. Because Luther wrote in a stimulating, polemical fashion and addressed himself to every level of society, he was the most widely read author of his age and has profoundly influenced the attitudes and ideas of generations to come, especially the Lutheran clergy.[59]

During his lifetime, Luther successfully campaigned against the Jews in Saxony, Brandenburg, and Silesia. Hoping that the violence of Luther's works would cause readers to reject them, the Strasbourg city council had initially refused to forbid the sale of Luther's anti-Jewish books, despite the pleas of Josel of Rosheim. But the city fathers were persuaded to institute a ban when a Lutheran pastor in nearby Hochfelden sermonized that his parishioners should murder the Jews.[60]

Luther's influence intensified after his death. Lutherans rioted against and expelled Jews from German Lutheran states through the 1580s.[61] At the end of the sixteenth century, both the Catholic Elector of Trier and the Protestant Duke of Brunswick set out to exterminate both witches and Jews. Some contemporary writers justified such activities. In 1589 the Jews of Halberstadt were charged with being "a peculiar vermin and a people that is insufferable among the Christians."[62] In 1596, Laurentius Fabricius, a professor of Hebrew at the University of Wittenberg, took up Luther's antisemitic ideas, arguing that "the Lord came specially for them, . . . but they made themselves unworthy of eternal life. . . . They . . . opened their mouths and all their senses to the Devil who filled them with all the lies, impiety, and blasphemy." Fabricius, like Luther, associated the Jews and their rejection of Christ with the devil. He wrote that Judaism "is no more the name of God. [It] was now turned into . . . 'dung,' stinking animal excrement, which Satan has set before the blind Jews to drink and eat, so as to make those who were nauseated by the dishes of divine mysteries sated with the most stinking excrements."[63] The Dominican friar turned heretic

Giordano Bruno (d. 1600) called the Jews "such a pestilential and leprous species, and one so dangerous to the public, that they deserve to be exterminated before birth."[64]

A Lutheran pamphlet of 1602 may be the modern starting point of the old myth of the Wandering Jew. A Lutheran minister named Paulus von Eitzen, who had studied with Luther in Wittenberg, met a bearded Jew named Ahasverus in a Hamburg church in 1542. Ahasverus was said to be the Jew who had mocked Jesus on the way to Calvary and was cursed with wandering and suffering until the Last Judgment. The booklet stated that the Jews were forever punished and that Christ himself denied them salvation because they refused to believe.[65] The publication was immediately popular and ran through almost 50 editions within a few years. The German term *Ewiger Jude*, or "Eternal Jew" (Wandering Jew), made its first appearance in 1694.[66]

Johannes Buxtorf the Elder (d. 1629) was a German Protestant scholar at the University of Basle. Although he published several volumes on the Hebrew language and Bible, his *Juden Schuel*, or *Synagoga Judaica* (1603), accused the Jews of deceptions and argued that Judaism was confused and disorderly. Ridiculing the notion of a Jewish Messiah, Buxtorf contrasted the carnal reign of the Jewish Messiah with the spirituality of the Christian Messiah, Jesus Christ.[67] He alleged that Jews conspired obscenely to corrupt the very food that Christians ate: "they defile and corrupt this meat with the urine of their sons and daughters" so that Christians would suffer a deadly curse.[68]

Luther's ideas were again disseminated in 1612 when Vincent Fettmilch reprinted "On the Jews and Their Lies" in order to stir up hatred against the Frankfurt Jews, whom he scapegoated for the depression of the Frankfurt cloth guilds, even though Dutch and English competition were probably responsible. In 1614 the people of Frankfurt rioted against the Jewish ghetto, where Jewish residences were already marked by insulting signs of pigs and scorpions. Nearly 3,000 Jews were killed and the rest expropriated and expelled.[69] Although the leaders of the pogrom were executed, the authorities convicted them of trying to overthrow traditional authority rather than of any offenses against the Jews.[70]

During the Thirty Years' War, 1618–48 (the Peace of Westphalia, which ended the war, essentially concluded the Reformation by recognizing both Protestantism and Catholicism, the rights of the rulers to choose the religion of the ruled, and the division of the Germanies that lasted until the time of Napoleon), which devastated large areas of central Europe, the Jews had been allowed into cities, reversing a long-standing trend toward dispersal in the countryside. Some Jews did well as suppliers of war materials to the contending armies and afterward some Jews, known as court Jews (*Hoffuden*), became financiers for the many extravagant German rulers. Nevertheless, the vast majority of Jews, many of them immigrants from troubled eastern Europe, lived poverty-stricken lives in the ghettos.

Luther's ideas and feelings about Jews and Judaism served as a basis for the essentially anti-Jewish worldview of many German Lutherans well into the twentieth century. (Luther had no equivalent in England, France, Italy, or Spain.) Most Germans were attracted by Luther's nationalistic views that German values were superior to alien, unGerman ideas and that government should hold great authority in the secular sphere.[71] Although Adolf Hitler never acknowledged that the "Final Solution to the Jewish Problem" was based on Luther's ideas, Hitler admired Luther and was quite aware of his antisemitism. Hitler's government followed Luther's program for dealing with the Jews very closely. In 1920, a member of a Bavariarn Guard unit—"a thoroughly well-meaning and honorable young man"—sent his "Recommendations for the Solution to the Jewish Question" to the Bavarian Minister–President: Jews were to be "rounded up" and "transported to concentration camps," Jews who resisted were to be killed and their property seized, Christians who sympathized with Jews were to be treated like Jews, any Jews who survived the concentration camps were to be deported to Palestine abandoning their property, returning to Germany meant death.[72]

Luther's name and antisemitic ideas were invoked on nearly every page of Dietrich Eckart's *Bolshevism from Moses to Lenin: A Dialogue between Adolf Hitler and Me* and several passages from Eckart's "Das ist der Jude!" quote Luther's *Von die Juden und Ihren Lügen* word for word. Just about every anti-Jewish book printed during the Third Reich, such as Theodor Fritsch's 1933 *Handbuch der Judenfrage*, contained citations to and quotations from Luther, among others, just as Eckart's book did. Luther's image appeared on iconic postcards during the Third Reich; in 1933, for instance, commemorating the four hundred and fiftieth anniversary of Luther's birth and in 1939 supporting the Nazi Winterhilfswerk des deutschen Volkes, winter relief program of the German People, with the inscription stating, "I seek nothing for myself, but all for Germany's success and prosperity." His photograph is to be found in the book *Antisemitismus der Welt in Wort und Bild*, an extremely popular book in Germany during the 1930s. Allusions to and quotations from Luther appeared in Streicher's *Der Stürmer* dozens of times in the 12-year Third Reich, 12 times alone between March 1939 and May 1941.

In 1933, Nazi Foreign Minister Konstantin von Neurath, celebrating the four hundred and fiftieth anniversary of Luther's birth, told an important audience at the state celebration (Reich President Hindenburg was there, among others) that "the mighty development of the spiritual life of the

German *Volk*, even outside the sphere of the religious, is unthinkable without Luther." Another speaker was the charismatic Hans Schemm, head of the National-Socialist Teachers' League and *Gauleiter* of Bayreuth and Bavarian minister of education and culture. Schemm—notable for his slogan "Our religion is Christ, our politics Fatherland!" and his pastoral speeches that often ended with Luther's hymn, "A Mighty Fortress Is Our God"—on the occasion of the *Deutsche Luthertag* sanctified antisemitism: "The older and more experienced he became, the less he could understand . . . the Jew. His engagement against the decomposing Jewish spirit is clearly evident . . . from his writing against the Jews . . . .[73] Elsewhere in Germany during this period, Hans Hinkel, later an influential member of Goebbels' Reich Chamber of Culture and Propaganda Ministry, where he headed the Jewish section, said that "with Luther, the revolution of German blood and feeling against alien elements of the *Volk* was begun. . . . To continue and complete his Protestantism, nationalism must make the picture of Luther, of a German fighter, live as an example above the barriers of confession for all German blood comrades."[74]

In 1935, Hitler's second in command, Hermann Goering, informed the Prussian State Council that Germany was "the leading Protestant country, from which the ideas and beliefs of Luther flowed over the world." (Goering had married his second wife in a Lutheran service and indicated that he would baptize his daughter in the Lutheran sacrament.)[75] From segregation and loss of rights, to expropriation and mass murder, the Nazi Final Solution proceeded apace with Luther's suggestions. In 1940, an admiring Heinrich Himmler told his friend Kersten, "what Luther said and wrote about the Jews. No judgment could be sharper."[76] It reminds us of Luther's "sharp mercy" mentioned above: German rulers should "cut, saw, and burn flesh, veins, bone, and marrow. . . . Burn down their synagogues, . . . force them to work, and deal harshly with them. [They are a] people possessed . . . ."[77]

Julius Streicher referred to Luther in his own defense on the stand at the Nuremberg War Crimes Trials. "Dr. Martin Luther," Streicher speculated, "would very probably sit in my place in the defendants' dock today [29 April 1946], if this book had been taken into consideration by the Prosecution. In the book, *The Jews and Their Lies*, Dr. Martin Luther writes the Jews are a serpents brood and one should burn down their synagogues and destroy them."[78]

The Nazis did not have to lie about Luther. He not only stood beside them, "rifle at the ready," he articulated and elaborated centuries of theological antisemitism into an especially virulent configuration that "prevented any large-scale mobilization of concern for the Jews" during the Holocaust.[79] On 17 December 1941, seven Lutheran regional-church confederations issued a joint statement that indicated their agreement with the policy of the Third Reich in identifying the Jews with a star, "since after his bitter experience Luther had already suggested preventive measures against the Jews and their expulsion from German territory. From the crucifixion of Christ to today the Jews have fought against Christianity . . . ."[80] Even Hitler's Lutheran opponents, the pastors Martin Niemöller and Dietrich Bonhöffer, were caught up in traditional Lutheran antisemitism.

After the war the German Lutheran Churches publicly stated that they felt some responsibility for their feelings of responsibility for the Final Solution. At Stuttgart in October 1945, the Lutheran Confessing Church proclaimed to the world a "confession of guilt."[81] In 1948, the National Brethren Council of the Protestant Church in Germany admitted that "The Churches had forgotten what Israel really is, and no longer loved the Jews."[82]

The rabbi and scholar Reinhold Lewin—who was deported in the 1940s along with his family to their deaths by the Nazis when the U.S. Embassy denied his request for visas for him and his family—made what may be the clearest analysis of Luther's long-term influence.[83] In 1911, Lewin wrote,

> [Luther] profoundly believed that the Jews acted only for the benefit of their own religion and against the interests of Christianity. He expounded this position . . . as a war to the death, and he wrote as a religious fanatic himself. The seeds of Jew hatred that he planted were . . . not forgotten; on the contrary, they continued to spring to life in future centuries. For whoever wrote against the Jews for whatever reason believed that he had the right to justify himself by triumphantly referring to Luther.[84]

Luther's ideas and activities demonstrate how traditional antisemitism affects modern state-sponsored pogroms. Luther, in part because he needed the Lutheran princes' protection, sanctioned the power of the state over that of the Christian Church, with its inherent inhibition against directly ordering the murder of Jews. The Nazis would not have approved of Luther's early focus on converting the Jews. But his scatological rhetoric, his violent policies, and his deadly conclusions about the Jews were nearly identical to those of the Nazi regime and virtually the same as Hitler's Final Solution itself. (Obsessed with the Jewish problem, in 1916 Houston Stewart Chamberlain quoted Luther when he concluded what kind of man Germany needed to save Germany: "a brave man with the heart of a lion to write the truth.")[85]

By the end of the eighteenth century, Christian Europe conceived, developed, and gave birth to Jewish emancipation. But the process of Jewish enfranchisement was not easy. Its most powerful foes—no matter how rationalist or secular their arguments seemed—relied, consciously or unconsciously, on the traditional form and content of religious antisemitism.[86] Modeling on Martin Luther, professor of Hebrew at Heidelberg, Johann Andreas Eisenmenger, wrote *Entdecktes Judenthum,* that is, *Judaism Exposed: Or a Thorough and True Account of the Way in Which the Stiff-Necked Jews Frightfully Blaspheme and Dishonor the Holy Trinity, Revile the Holy Mother of Christ, Mockingly Criticize the New Testament, the Apostles, and the Christian Religion, and Despise and Curse to the Uttermost Extreme the Whole of Christianity.* Reflecting centuries-old Catholic anti-Jewish defamations and more recent Lutheran ones, Eisenmenger attacked the Talmud, rabbinic Judaism, and Jews as blasphemous and dangerous. The kernel of truth in Eisenmenger's analysis is that Jews did pursue an ethical dualism, regarding themselves as different from Gentiles; and they dreamt of a messianic future. But, as Jacob Katz points out, so did Christians.[87] His two huge volumes sought to convince Christians that the Jewish holy books lay behind Jewish ritual-murder and blood-libel, that Jewish poisoning of wells caused the great plague of the fourteenth century, and that Judaism commanded Jews to lie to Gentiles and to be disloyal to the Christian state.[88] Eisenmenger maintained that Jews' "hatred of Christians is just as great now as it might have been at any time past."[89] Eisenmenger described the alleged Jewish murder of the two-year-old Simon of Trent:

> An old Jew named Moses took the child on his lap, undressed him, and stuffed a cloth in his mouth so that he could not cry out. The other held his hands and feet. The said Moses made a wound with a knife in the right cheek and cut out a small piece of flesh. Those standing around caught up the blood, and each cut out a small piece of flesh until the wound had become the size of an egg. This they did on other parts of his body. Then they stretched out the hands and arms like a crucifix and stuck needles into the half-dead body . . . : "Let us kill him, just as we did the God of the Christians, Jesus, who is nothing. Thus must all our enemies perish." . . . The Jews [of our day] do not scruple to kill a Christian and . . . it must be permitted to them, if only it can be done conveniently, secretly, and without danger.[90]

Frank Manuel calls Eisenmenger's book the birth of "Scientific Judaeophobia."[91] The Frenchmen Poujol and Louis Rupert; the Dane C.F. Schmidt-Phiseldeck; the Germans Friedrich Rühs, Sebastian Brunner, Albert Wiesinger; the Austrian writers for the *Wiener Kirchenzeitung* (the *Vienna Church News*) and the Austrian priest August Rohling; and twentieth-century National-Socialists based their own antisemitic polemic on Eisenmenger's work.

The most influential German encyclopedia of the early eighteenth century, *Grosses Vollständiges Universal Lexicon*, followed Luther's lead in its treatment of the Jews.[92] Its article on Jews contended that once the Jews rejected Jesus as the Christ, they became Satan's people. "And how often have they slaughtered Christian children, crucified them, battered them to pieces! They are the worst thieves, and betrayal is their distinctive insignia. . . . Yes, God has marked them in their very natures. . . . [T]hey killed the Son of God, and crucified the Lord of Glory, whose blood still weighs upon them . . . ."[93]

Most Christians held to traditional antisemitism during the Enlightenment, among them the most "rational" of the philosophers, such as Denis Diderot and Voltaire. The *St. Matthew Passion* of Johann Sebastian Bach, a self-conscious Lutheran, intones "Let him be crucified!" The music at this point strikes a tritone, called in the Middle Ages "the Devil in music."[94] The chorus then demands seven times, "Let his blood come over us and our children."[95]Bach's *St. John Passion* also follows the Gospel's defense of Pilate's innocence and revelation of the Jews' eternal responsibility for Jesus' crucifixion. Mentioning "the Jews" a dozen times, Bach has them cry out, "Away with him, away with him, crucify him!"[96] Although Bach's texts are limited to the actual words of the Gospels, this music of his reminded its listeners of the attribution of Jesus' crucifixion to the Jews, historically inaccurate but psychologically satisfying and theologically necessary.

The Oberammergau passion play, first performed in 1633, Europe's most popular passion play, staged more than 3,000 times,[97] employs the language of Matthew's Gospel to indict the Jews again and again for the crucifixion of Christ, for calling down "His blood . . . upon us and upon our children." In 1942, Hitler ordered that "It is vital that the Passion Play be continued at Oberammergau; for never has the menace of Jewry been so convincingly portrayed . . . ."[98] (Even "the great European," Gottfried Wilhelm Leibniz, tempered his friendship with Jews by his anti-Jewish stereotypes.)[99]

Although anti-Jewish sentiments still predominated in seventeenth- and eighteenth-century Europe, a few Christian thinkers felt akin to biblical Judaism: poets, painters, and playwrights, such as Milton, Rembrandt, and Racine; others were religious figures such as the Swede Anders Pedersson, the Dutchman Petrus Serrarius, the Huguenot living in Holland Pierre Jurieu, the Dane Holger Pauli, the German Heinrich Küster, the Anglo-Irishman John Toland. The Augsburger Johann Peter Spaeth converted to Judaism and took the name Moses Germanus.[100] Playwrights treated Jews positively but with reservations. Christoph Wieland confessed his sympathy for the God-cursed Jews, whom Christians victimized by contempt, injustice, and greed. "We persecute them because God has punished them for their sins with blindness so that they, for the sake of their redemption, cannot believe what we believe . . . ."[101] In 1746, Christian Fürchtegott Gellert, the so-called Educator of Germany, *Praeceptor Germaniae*, made the point that good Jews did exist. Not all of them were usurers and Antichrists.[102] Although antisemitism may have colored his confession that a perfect Jew would have made him forget that the person was a Jew, Gotthold Ephraim Lessing wrote two plays sympathetic to Jews.[103] He wrote *Die Juden* in 1749, 11 years after the Court Jew (Court Jews acted as army provisioners, moneylenders, administrators, and tax collectors for the princes of Germany, enhancing their power at the expense of the burghers, and often flaunted their wealth and Christian mistresses)[104] Joseph Süss Oppenheimer (Jud Süss), the most influential official of Würtemberg, was hanged in Stuttgart. For many Christians, Oppenheimer had symbolized the "illicit" authority of a Jew over Christians and was executed as a demonic Jewish upstart.[105] Portraying the Jewish protagonist as a virtuous man, *The Jews* ridiculed the idea of Christian triumphalism. Lessing's play, *Nathan the Wise* (1779), was based on Lessing's friendship with Moses Mendelssohn, son of a Torah scribe.[106] The play's most important Christian character, a knight Templar, is transformed from a Jew-hater to a Jew-lover who discovers that his own sister was raised by the Jew, Nathan. Lessing describes Nathan, a banker, as "A Jew to whom . . . his God had given, of all the goods of earth, in fullest measure, . . . wealth [and] wisdom. . . . How free his mind of prejudice; how open his heart to every virtue, how attuned to every beauty." Lessing questioned parochial German nationalism, asking, "Do we consist only of what our nation is? What is a nation? Are Jew and Christian, Jew and Christian first, or human beings first?" Lessing noted that the brilliant German philosopher Immanuel Kant "cannot bear a hero from this people. . . . This is how divinely severe our philosophy is in its prejudices despite all its tolerance and impartiality."[107] Johann Gottfried von Herder, a pastor, theologian, and philosopher who was a contemporary of Goethe's at Weimar and Kant's student, praised Lessing's *Nathan the Wise* for its teaching of "toleration of human beings, of religions, and of nations." Its message, he noted, was "be human beings!"[108] Mendelssohn wrote to Herder, "Moses, the human being, is writing to Herder, the human being, and not the Jew to the Christian Court preacher."[109]

Herder wrote often on the Jews, criticizing the state as "barbaric" for not granting Jews legal equality. Although he felt that the Jewish people were terribly lacking in skills for statecraft, warfare, science, and art, Herder praised the ancient Jews and their ethics as embodied in the Torah, Talmud, and Midrash. As long as the Jews were treated like slaves, they could not become fellow citizens. As long as spendthrifts need funds, Jews would be involved in usury. As long as Christians mistreat them, Jews will not contribute to the best interests of the state. Both Christians and Jews need to be reformed.[110]

But Herder, disavowing "animal racism," nevertheless adhered to a kind of spiritual racism.[111] The Jews, he wrote, "has for millennia, and almost since its origin, been a parasitic plant on the stems of other nations [and] nowhere yearns for its own honor, a domicile of its own, or a fatherland."[112] Herder's antagonism toward Jews provided a moral ground for many German antisemites of his day and led to his idealization in Nazi Germany.

Dohm and Jewish Emancipation

The Prussian king Friedrich-Wilhelm I instructed his son, the future Frederick the Great: "These [Jews living in Prussia without royal permission] you must chase out of the country for the Jews are locusts in a country and ruin the Christians. . . . You must squeeze them for they betrayed Jesus Christ, and must not trust them for the most honest Jew is an arch-traitor and a rogue."[113] Frederick the Great, who severely exploited the Jews economically, on hearing that the Jews planned to convert *en masse*, exclaimed, "I hope they won't do such a devilish thing!"[114]

In 1781, the forty-first year of Frederick the Great's reign and the year Lessing died, at the urging of his friend Moses Mendelssohn, Christian Wilhelm von Dohm, historian, economist, the former editor of a Göttingen journal, and the busy archivist for the Prussian *Kriegsrat*, or War Council, published a book that favored Jewish emancipation[115] based on the rationalization of the Prussian state, where all citizens, including Jews, would be productive and happy members of society.[116] Dohm hinged his demand that Jews be granted civil rights on the requirement that Jews "must cease to be Jews."[117] He suggested that Christian hostility to Jews had conditioned them not to "feel humanly." He also claimed that the Jews were "more morally corrupt than other nations," they were "guilty of a proportionately greater number of crimes than the Christians," "their character in general inclines more toward usury and fraud," and "their religious prejudice is more antisocial and clannish" than Christians. Events have conspired, he wrote, "to choke every sense of honor in [the Jew's] heart."[118]

Other Germans, such as August Krämer from Regensburg and the pastor Johann Ewald, took a position similar to Dohm's. Professor Alexander Lips of Erlangen observed that "we [Christians] inoculate the child with hatred of Jews; at later ages we diligently nourish that hatred; but we do not condemn the causes that engender it in us and the springs of it, which lie in ourselves, our own hatred, and our own spirit of exclusiveness. . . . [Christianity must practice humanity and justice in order] to hold out our hand to our long-forgotten brothers."[119]

Austrian emperor Joseph II's Edict of Toleration of 1782 permitted Jews to enter the Austrian economy, remove stigmatized clothing, have Christian servants, appear in public on Sunday, and wear swords. But the decree did not allow Jews to own land or shops outside the ghetto or to have public synagogues in Vienna. They were still required to pay a special Jew tax, to speak German in their business transactions, and to take German names. Pope Pius VI and the Austrian bishops opposed all the rights that the throne granted the Jews.[120]

Goethe: A German Genius and the Jews

Johann von Goethe, perhaps Germany's greatest literary figure, felt ambivalent about Jews: "The most serious and most disastrous consequences are to be expected [from Jewish emancipation] all ethical feelings within families, feelings which rest entirely on religious principles, will be endangered . . . ."[121] Goethe respected his Jewish intellectual readers, but he expressed contempt for the mass of Jews, who were in no way like their biblical forebears.[122] He called Judaism "ancient nonsense" and the modern Jewish language, Yiddish, the vehicle of a rabbinic, anti-Christian fanaticism.[123] "We tolerate no Jew among us, because how could we grant him participation at the highest level of [Christian] culture, whose very origin and tradition he denies."[124] While denying that he hated the Jews, Goethe opposed Jewish emancipation and equal rights, and he ridiculed those German-Christians who supported Jewish liberation. He approved of the anti-Jewish behavior of the local prince-primate "for treating this people as they are and as they will remain for a while," and he disapproved of a law permitting the marriage of Jews and Christians. "I think the Superintendent should rather resign his office than tolerate the marriage of a Jewess in the name of Holy Trinity [in] contempt of the religious feelings of the people . . . ."[125] Although he found it difficult to shake off the religious antisemitism that permeated most of German society,[126] at times he seemed able to rise above it: "My contempt [for the Jews] was more the reflection of the Christian men and women surrounding me. Only later, when I made the acquaintance of many intellectually gifted, sensitive men of this race, did respect come to join the admiration I cherish for the people who created the Bible and for the poet who sang the Song of Songs."[127]

Immanuel Kant: Only as Christians Could Jews Obtain Moral Standing

Immanuel Kant, the greatest moral authority of his age, advocated the eradication of Judaism and the de-judaization of Christianity. He argued that formalistic, legalistic Judaism could not compare with Christian love, which emphasized the "moral intention in an act."[128] He considered Judaism a set of laws and a ritualistic practice that was merely "a mechanical cult," grounded in that abomination of reason called blind faith.[129] Judaism contained no authentic morality; even the Ten Commandments are followed in rote obedience, not genuine ethical concern.[130] Although Kant had Jewish students and was friendly with Mendelssohn, he saw Jews only through a traditional Christian lens. Believing that Jews would always be a social menace, Kant told a friend that "As long as the Jews are Jews and allow themselves to be circumcised, they never will become more useful than harmful to civil society. They are now vampires of society."[131] "Only through self-euthanasia, publicly accepting the religion of Jesus,"[132] could Jews become fruitful members of Christian society and obtain political rights.[133] Kant also wrote, "The Euthanasia of Judaism is the pure religion of morality, with the abandonment of all old doctrines, of which some must still be retained in Christianity (as the messianic belief)." Poliakov terms Kant's views on Judaism as a "metaphysical way of crying: 'Death to the Jews!' "[134]

Fichte: Radical Antisemite

The anti-Jewish ideas of Johann Gottlieb Fichte were more radical than the peaceful euthanasia of Judaism that his teacher Immanuel Kant had proposed. A former student of theology who had prepared for a career in the Church, Fichte stirred a hate campaign against the Jews based in great part on religious differences between them and Christians. Part of his 1793 essay on the French Revolution argued: "Within all the nations of Europe the Jews comprise a hostile state at perpetual war with their host nations [and] founded on the hatred of the whole human race. . . . The Jewish nation excludes itself from our [German-Christian culture] by the most binding element of mankind—religion."[135] He criticized the French for their "loving toleration" of "those who do not believe in Jesus Christ." Fichte considered Jews so evil that he did not want them "to believe in Jesus Christ."[136] He argued further—and most notoriously—that Jews should not be awarded civil rights until one night Christians "chop off all their heads and replace them with new ones, in which there would not be one single Jewish idea."[137] Fichte may not have been advocating physical annihilation of the German-Jews, but he clearly wanted to destroy Jewishness. Expulsion was another solution: "And then, I see no other way to protect ourselves from the Jews, except to conquer their Promised Land for them and send all of them there."[138] Fichte's *Addresses to the German Nation* asserted that only Germans could be genuine Christians; only Germans were qualified to detect "the seed of truth and life of authentic Christianity."[139] (Fichte later claimed not to have really known Jews at this time, praised the ancient Jews, argued that Jews must possess human rights, and denied that his heart contained the least bit of "the poisonous breath of intolerance.")

Fichte was not alone in using the language of violence with regard to the Jews. Lesser-known nineteenth-century writers applied metaphors of torture and murder to Jews: "beat to death," "exterminate the gnawing worms," "circumcision and castration," "parasite extermination."[140] The twentieth century would see the actualization of this violent rhetoric. The National-Socialists and their collaborators carried out Fichte's proposals but reversed the order. The Nazis first attempted to expel the Jews, but when the rest of the world refused to cooperate with this plan, the Nazis then executed the remaining alternative, and "chopped off all their heads." Fichte's explosive combination of Christian antisemitism and German nationalism would become a fundamental ideological stimulus of the Holocaust.

Nineteenth-Century Germany

In few other European countries were the issues surrounding Jewish emancipation so thoroughly and painfully debated as in the Germanies, a welter of different states with inconsistent legislation on the Jews. Although Germany became increasingly secularized during the nineteenth century, German society and culture were essentially Christian. German political, economic, and scientific groups defined themselves as essentially Christian.[141] Even consciously anti-ecclesiastical Christians still "retained allegiance to Christian notions and doctrines,"[142] including Christian antisemitism.

Remaining attached to the anti-Jewish Christian worldview, myths, and symbols that they had absorbed in their youth and that they shared with most of their countrymen, most German thinkers, theologians, and politicians believed that the "Jewish Question" was crucially important.[143] They considered the Jews the primary enemies of Christianity and of the German nation. Most Germans regarded the Jewish denial of Jesus as signifying the

Jewish rejection of all of civilization and humanity. German-Christians asserted their own humanity by opposing the Jewish apostles of Satan.[144] All shades of the political spectrum held that the Jews were morally corrupt deicides undeserving full civil rights and political equality. A Jew could never be fully German.[145] Indeed, Jews were hardly human, symbolizing "the sinister, superstitious, backward, irrational, medieval, stubborn, filthy, ultimate subhuman."[146]

Unlike the relatively secularized French governments, German governments presumed themselves to embody Christian values.[147] The surest way to insult or destroy a political opponent was to call him a Jew. For most German conservatives, liberals, and socialists the Jew was still the Antichrist. As Robert Alter has recently put it, "A deep kinship between left and right was expressed in the loaded opposition between *Deutschtum* and *Judentum* through which both sides of the political spectrum sought national self-definition."[148] Conservative Germans assumed that the Jews represented the disintegration of modern life, sometimes surreptitiously, other times obviously, and that Jewish greed, materialism, and monopoly were ruining Germany's traditional social, religious, and cultural life.[149] Conservatives reacted against the forces of change, represented by the Jews, by supporting a German-Christian state that justified itself morally by its grounding in the Christian religion and its excluding Jews.[150] The Christian state was to be a recapitulation of the medieval *societas Christiana.*

Some liberals supported the Jewish desire for emancipation from centuries of legal and social disabilities. These liberal German-Christians, like their French and English counterparts, anticipated that the traits they disliked in the Jews would disappear once Jews were "liberated" and assimilated. That is, the emancipation "contract" that Jews were expected to enter into with their Christian "hosts" required that in return for obtaining political rights and entry into polite Christian society, Jews would have to cease being Jews, or they would continue to be shunned.[151] Even when German-Jews bought into the "social contract" the liberals demanded—leave the ghetto, engage in a variety of occupations, drop the Yiddish language, reach out to their Christian neighbors—the German governments held back granting Jews equal rights. The Prussian government, for instance, feared that a reformed Judaism would be too attractive to Christians.[152]

Most liberal thinkers felt that Jews deserved the rights of citizenship but had to convert to Christianity so that they would be like all the other Germans. Heinrich Heine had called conversion "an admission ticket to western culture." But to Christian-Germans, Heine, although a maestro of German language and literature, remained a Jew and an outsider.[153]

A lingering fear haunted German liberals that Jews would persist in their "alien" ways and destroy the new "liberal" democratic society emerging in the nineteenth century. German liberals may have believed in a society blessed with legal and fiscal equality and with political rights for all; but these liberals were not contemplating individual rights that would guarantee Jews the freedom to express and enjoy their Jewishness openly.[154]

Germans may have feared the achievement of the very goal they claimed to seek. For the more Jews assimilated—as the *conversos* in fifteenth-century Spain had—the more they resembled "regular" Germans, the more of a threat they were to German-Christian society once anti-Jewish restrictions were removed. Jews entered German society when the Enlightenment began to take hold, but a century later, German society abandoned them to their rational humanism. Germany became a nation in opposition to "unGerman" Enlightenment values.[155] Because of cultural and historical experiences, both German-Christians and German-Jews had strong family values, were devoted to hard work and their religion, respected education, were transnational, were disliked by the rest of the world, alternated between servility and arrogance, were indispensable and troublesome, aggressive, self-pitying, vilified, musically talented, and intellectually daring—but with completely different attitudes toward violence.[156] In the United States, *American* Jews were distinguished from *American* Christians, but in Germany the distinction was between *Jews* and *Germans*, between *Judentum* and *Deutschtum*. German-Jews were guests of the Christian-German *Wirtsvolk*, host people; allowed to exist in Germany on the basis of the native Germans' goodwill. Jews were stepchildren who had to be twice as good, *Stiefkinder müssen doppelt so brav sein*. Judaism was the religion of Jews, Christianity was the religion of the German nation, of the German *Volk*.[157]

Even German socialists, hostile to Christianity, were "imbrued with the basic thought patterns of that Christian theology which [they] sought so fiercely to eradicate."[158] Socialists associated Jews with exploitative capitalism and argued that Jews caused the moral degeneration of bourgeois society. Whereas biblical Jews were seen as crucifying Jesus, contemporary Jews were crucifying Christians on a cross of gold.

The Wars of Liberation and Their Aftermath

When Napoleon conquered the Germanies he introduced the French Revolutionary reforms that initiated Jewish emancipation. The ensuing nationalistic wars against the French encouraged German regimes to grant concessions to the Jews to ensure their cooperation. But once Napoleon was defeated in 1814–15, all across Europe reactions against the democratization of European societies set in and the old restrictions against the Jews were restored. In Frankfurt, the Jew Judah Baruch lost his job as police actuary even though his father had represented the Jews of Frankfurt at the Congress of Vienna. He wrote, "We were set free from the purgatory of French rule so that we might gain redemption, not into paradise, but into hell." In 1818, Judah Baruch converted to Christianity as Ludwig Börne.[159]

Even before its defeat by Napoleon, Prussia, the leading German state, was experiencing a reform movement prompted by the French Revolution. Baron Karl vom Stein, the patriotic and religious leader of Protestant Germany, was the leading German statesman and reformer of the Napoleonic period. Although he was never personally well disposed toward Jews, he accepted them into the Prussian state when the state needed them. Publicly, Stein took the position that Jews could become acceptable only by shedding their Jewishness and becoming Christians. Privately, he expressed a desire to expel them from Germany. In 1816, Wilhelm von Humboldt wrote to his wife that Stein would "populate the northern coast of Africa with them."[160]

The liberal politician, Prince Karl von Hardenberg, appointed Prussian Chancellor a second time in 1810, became an ardent advocate of Jewish liberty and equality, although his sympathies did not extend to the Polish Jews who came under Prussian control after the Napoleonic wars and who comprised the majority of Jews living in Prussia.[161] A rationalist, he wanted to involve all elements of Prussian society in public life; Jewish liberation and participation provided an important part of his reforms, which he hoped ultimately would spread to all Germany and supply the moral improvement Jews required. Hardenberg's program was offered conditional support by Prussian liberals, who felt that limited freedom for Jews would, as Baron von Schrötter (vom Stein's right-hand man) argued, "undermine their nationality, destroy it, and thus gradually induce them no longer to aim at forming a 'State within a State.' "[162] Wilhelm von Humboldt, who followed Stein into office and supported reforms involving Jews, also harbored significant reservations. If the Jews refused to accept full citizenship and assimilate completely into Prussian society, both Humboldt and Schrötter felt that they should be expelled.[163]

Important conservative figures, such as King Frederick Wilhem II, his successor Frederick Wilhelm III, and Baron Friedrich von der Marwitz, opposed Jewish emancipation in principle. They feared that traditional Prussia would be turned into a modern *Judenstaat* (a rationalized state dominated by merchants, bureaucrats, and Jews, to the enormous detriment of the nobles and the peasantry).[164] Marwitz and the Prussian nobility expressed their attitude toward the Jews in 1811: "The Jews if they are really true to their faith are the necessary enemies of every existing state . . . and have the mass of cash in their hands."[165] Nevertheless, under pressure of the approaching War for Liberation against the French, Prussia reluctantly enacted Hardenberg's series of reforms, including the emancipation of the serfs, several military reforms, and the Edict of 1812, concerning the Jews' rights of citizenship. Hardenberg intended the edict to treat Jews as a religious community like any other, and he expressed his political intentions in four words, "same rights, same duties."[166]

But the edict granted only limited citizenship to Prussia's Jews, who were precluded from entering state service or teaching. In 1817, Hardenberg acceded to the conservatives, who held that the Jews should not be allowed to participate in government service nor in the national assemblies "because they obviously corrupt the spirit of both."[167]

After the Napoleonic era, the popes set the tone for Catholics in Germany and elsewhere. Pope Pius VII decreed that the freedom Jews had been granted under the French should be terminated.[168] In 1832, Pope Gregory XVI condemned the liberal principle of "freedom of conscience [and] emancipation" as an "absurdity." In 1860, Pius IX approved a declaration that repeated the traditional Catholic arguments for the inferiority of contemporary Jews.[169] In 1864, his Syllabus of Errors attacked the secularizing tendencies of the modern world as "a war . . . waged against the Catholic Church." In a postscript, the pope stated that all these tendencies only serve to strengthen "the synagogue of Satan, which gathers its troops against the Church of Christ."[170]

Although roughly 400 newly emancipated Jews fought in the Prussian army, with nearly one-fifth earning the recently created Prussian Iron Cross medal, nationalistic German student organizations later expressed a high degree of antisemitism. Their leaders—Arndt, Jahn, Rühs, Fries, Oken, Luden—coupled traditional anti-Jewishness with a reborn nationalism fostered by both Lutheran and Catholic theology. At the Wartburg Festival of 1817, the German students burned books associated with Napoleon and freedom for Jews. (In 1820, reflecting his own people's history, Heine prophesied that "Where they burn books, there in the end they will burn people.") Two years later, the so-called HEP-HEP riots against Jews (HEP may refer to a Crusader slogan, "*Hierosolyma est perdita*," that is, Jerusalem is lost. These pogroms repeated in 1830, 1834, 1844, 1847–48) were economically and religiously motivated and widely supported all over Germany and several other parts of Europe.[171] Although a few Christians, such as some students and professors at Heidelberg, defended Jews and slowed the assault, the police often stood by without taking action. The army was finally called in because the authorities feared that the riots could spread to other targets besides the Jews. A broad spectrum of Germans, fearful that emancipated Jews would challenge Christian dominance at all levels of society, supported these anti-Jewish actions. The HEP riots were conspicuously religious. The rioters often carried flags with crosses or Judas dolls hanging by their necks, and the pogroms were often preceded by church sermons on Jewish ritual murder. Synagogues were destroyed and Jewish holy objects desecrated.[172] In Würzburg, where the HEP riots began on 3 August 1819, at a time when many Lutherans still adhered to the ritual-murder myth, university students shouted down all defenders of the Jews and read the following proclamation:

> Brothers in Christ!
> Arm yourselves with courage and strength against the enemies of our religion; it is time to crush the race of Christ-killers. . . . Down with them, before they crucify our priests, profane our sanctuaries, and destroy our temples. . . . Let us now execute the judgment which they have called down upon themselves, "Let His blood come over us and our children!" . . . Arise, in a holy cause. These Jews, who live here among us, are spreading like rapacious locusts and threaten all of Prussian Christendom with subversion. They are the children of those who once cried, "Crucify, crucify!"[173]

Rahel Levin, a prominent Jewish intellectual and social leader in Berlin, was shocked by the riots. She wrote to her brother that the Germans tormented, insulted, despised, and physically assaulted the Jews because Germans hypocritically claimed to love Christianity. "The Christian religion [and] the Middle Ages with its art, poetry, and outrages, [incite] the people to the only atrocity to which, reminded of the old experiences, it still permits itself to be incited."[174]

Most German-Christians found ideas of equality and humane treatment for Jews repellent.[175] In February 1832, Ludwig Börne (Judah Baruch) bitterly summed up the predicament of Jews who were making their way into German-Christian society: "Some reproach me with being a Jew, some praise me because of it, some pardon me for it, but all think of it."[176] The humanist Wilhelm von Humboldt was an exceptional spokesman for Jewish rights in Prussia and at the Congress of Vienna. (The Congress, which had granted all Christians civil and political rights, left to the individual German states the decision to grant such rights to Jews. The Jewish issue arose because the imperial cities, Frankfurt, Hamburg, Bremen, and Lübeck, had repealed Jewish rights, and the Jews appealed to the victorious Allies for redress.) As a youth, he knew several Jews, learned Hebrew, and argued strongly for equality and Jewish emancipation.[177] But Humboldt, like Kant, looked at Judaism as an inferior, legalistic religion that would soon be abandoned by the assimilated Jews in favor of "a loftier faith," as he put it.[178]

Humboldt's wife, Caroline, was a convinced antisemite who, despite having several Jewish friends, including Rahel Levin, wanted the Jews to disappear.

"[T]his would . . . be a gain for humanity; the Jews in their depravity, their usury, their inherited lack of courage, which springs from usury, are a stain upon humanity."[179] She felt the Jews controlled "all the money of the country" and called them the "stigmata of the human race."[180] In 1816, Caroline noted that "in fifty years the Jews would be exterminated, as Jews."[181]

Conservative military men, such as the theorist Karl von Clausewitz, the strategist and field marshal August von Gneisenau, and the old general Prince Gebhard von Blücher, perceived the Jews through lenses darkened by religious antisemitism. Like Fichte a member of the anti-Jewish Christian-German Table Society, Clausewitz observed that the "filthy German Jews, who swarm in dirt and misery like vermin, are the patricians of this country." (A fourth Prussian military figure, Gerhard von Scharnhorst, seemed to be genuinely philosemitic.)[182]

Romantic conservatives such as the brothers Grimm discovered the same negative hostility to Jews in German fairy tales as Achim von Arnim and Clemens Brentano found in German folk songs and tales. (Arnim and Brentano founded the Christian-German Dining Club, which barred "Jews, converted Jews, and the descendents of converted Jews" from membership. But Brentano's sister, Bettina von Arnim, championed the Jews' cause, and Clemens himself may have softened toward them in his old age.)[183]Although he admired ancient Judaism, Brentano hated contemporary Jews, calling them ugly, traitorous, and nasty. He told an audience in 1815, "The historic deformation of Judaism follows from a rebellion of this tribe against its flowering and fruit [Jesus]; they therefore have sprouted out into a stale, sickly, ugly weed."[184] Friedrich Schelling, the philosopher of Romanticism, visiting Berlin in 1787, criticized the Jews, among whom he found a "vermin" of unbearable young scholars.[185] The poet and critic Friedrich Schlegel, the German "knight of Romanticism," who glorified Catholicism in his work and married Moses Mendelssohn's daughter Dorothea (she converted) rejected full civil rights for Jews.[186] Pastor Friedrich Schleiermacher, a friend of Schlegel's, a religious revivalist and the "church father" of modern German Protestant theology, asserted that Judaism was nothing but "a system of universal immediate retribution," which was now totally lifeless, hanging and dying on a withered stem.[187]

Hegel and Schopenhauer: Great Philosophers, Narrow Minds

Georg Wilhelm Friedrich Hegel was educated to be a minister and in his early years followed triumphalistic Christian beliefs, arguing that

Christianity, with all its flaws, was superior to Judaism.[188] Like Kant, he was ignorant of the true nature of Judaism. Assuming that the Jews suffered throughout history because of their flawed, legalistic religion, which made them what they were, Hegel maintained that the Jews would be "unregenerable" until they lost their "Jewish consciousness."[189] Calling the Jewish prophets "impotent fanatics," he denounced the unspiritual materialism of the Jews, the anti-German nationalist implications of the Jewish idea of the Messiah, and the Jews' inherent inability to care for anyone but themselves—all traditional Christian criticisms. He accepted the Christian stereotype of the pharisees as legalistic oafs in contrast to compassionate Christians imbued with spiritual beauty.[190] "The Jewish multitude was bound to wreck [Jesus'] attempt to give them something divine; something great cannot make its home in a dunghill [nor] the infinite spirit . . . in the prison of a Jewish soul."[191] In his *Philosophy of Religion*, Hegel later changed his mind, calling Judaism "a religion of sublimity" and asserting that the Jews deserved human, as well as civil, rights.[192]

Many of Hegel's students were hostile to Jews and Judaism. Arnold Ruge[193] held that Judaism was responsible for Christianity's flaws and compared Jews to "the maggots in the cheese of Christianity,"[194] an analogy found in both Nathaniel Hawthorne (*The Marble Faun*) and Adolf Hitler (*Mein Kampf*).[195] David Friedrich Strauss claimed that contemporary Jews observed a defunct religion made obsolete by Christianity's higher ethical code. Ludwig Feuerbach denigrated Judaism as a narrow, nationalistic, egoistic religion of "human sacrifices." Jews, he felt, had to convert to Christianity before they could be absorbed into a free Christian Europe.[196] Bruno Bauer advocated the disappearance of both Christianity and Judaism, but he held that the unique Jewish identity, stubbornly preserved over centuries, was responsible for the suffering Christians had caused the Jews.[197]

Like Luther, Arthur Schopenhauer blamed the Jews for having gained control of whole nations through robbery and murder,[198] without their own nation, the Jews live "parasitically on other nations and their soil."[199] To admit Jews to the "administration and government of Christian countries," he wrote, "is absurd; they are and remain an alien, oriental people; they must therefore always be treated as aliens."[200] Because Jews have sinned against the Savior and World-Redeemer, they must wander homeless in foreign lands. Jews should marry Gentiles so that in the course of a century, only a very few Jews will remain, and the "chosen people" will no longer know where their original home was located. This desirable goal will not be met if the Jews obtain political rights."[201]

Mid-Century

Like Christian antisemites of prior centuries, Germans of every political stripe and class considered Judaism inferior to Christianity and argued either that Jews had to convert in order to become full citizens or that they were so soiled by their Jewishness that no amount of baptismal water could cleanse them. Germans felt that "the Jew was not one of us," that something was *wrong* with being Jewish. The young Chaim Weizmann, later a world-famous scholar, chemist, and first president of Israel, felt that the Jews were like a splinter in the eye, even if it were gold, it was still an incapacitating irritant.[202]

Mendelssohn, Börne, Heine, and Marx, four Jews who were converted to Christianity during this period, could not free themselves from the stigma of having been born as Jews.[203] As a ten-year-old, Felix Mendelssohn was called *Judenjunge*, "Jewboy" by the Prussian prince before he spat in Mendelssohn's face. German street children called Mendelssohn Jewboy before they attacked him and his sister.[204] Later in life, Mendelssohn used the term about himself. (Mozart's librettist, the Venetian Lorenzo da Ponte, was also called Jewboy.)[205] In 1832, the satirist, Friedrich von der Hagen (Cruciger), called Börne "a Jewish wolf and changeling, or one of the masks of the Wandering Jew [roving] around until the Last Judgment." Cruciger called Heine a Wandering Jew who reflected "the highest degree of blasphemy against God and the Cross." (In nineteenth-century Germany, the Wandering Jew was the Eternal Jew (*Ewige Jude*) who "came to symbolize the Antichrist and death.")[206] Heine himself wrote, "I much regret having been baptized. . . . Isn't it absurd how scarcely am I baptized than I am cried down as a Jew?"[207] Heine called Jewishness "an incurable malady."[208] Several writers attacked Marx, including Mikhail Bakunin, who called Marx "the legislator of the German-Jewish socialists" and observed that the evil Jewish world existed "in large part at the disposal of Marx on the one hand, and of Rothschild on the other." ("I am in no way an enemy or a detractor of the Jews," claimed Bakunin. But he also maintained that Jewish history "has imprinted on them a trait essentially mercantile and bourgeois, which means, taken as a nation, they are par excellence the exploiters of the work of others, and they have a horror and a natural fear of the masses of the people, whom, moreover, they hate . . . . They constitute a veritable power in Europe today.")[209] The early French socialist Proudhon argued that Marx and Rothschild both belonged to the same parasitic "race of leeches."[210] In letters in 1844 and 1845, Arnold Ruge saw Marx as "a skunk and a shameless Jew" and his socialism as a "community of atrocious Jewish souls."[211] None of his critics could hate Jews more than Marx himself, who created a mythical capitalist Judaism and then attacked it—an authentic antisemitic technique—along with calling his opponents "little Jewy" and Yid.[212]

The playwright Karl Gutzkow wrote that German prejudice against Jews was essentially "a prejudice of religion."[213] To be truly redeemed, the Jews had to abandon their Jewish identity and assimilate into the surrounding Christian community.[214]

Both before and after the failed democratic revolts against Europe's conservative regimes in 1848, Germans exhibited a multitude of prejudices against the Jews, who played an active part in the revolts. In 1841, the conservative Ernst Arndt saw the Jews behind the disturbing social changes then occurring in Germany: the Jews hated "folkdom and Christianity" and "labor[ed] untiringly to destroy and dissolve [sacred German] patriotism and fear of God."[215] Other conservatives regarded the Jews as modern incarnations of Judas. Socialists mixed religious antisemitism with the belief that the carnal Jews were materialistic capitalists. The liberals told Jews, "to reform, conform, or to depart."[216] The peasants were caught up in traditional Christian antisemitism. The churches, at best, supported the position that Jews had to convert; at worst, they considered the Jews Antichrists. Others mixed Christian and modern racism into their antisemitic ideology.

To discredit the 1848 revolutions, the conservatives scapegoated the Jews as Judases responsible for it. King Friedrich Wilhelm attacked "the existence and the influence of that despicable Jewish clique with its tail of silly and foolish yelpers[—]a misfortune for Prussia. [They] by word, letter, and picture daily [lay] the axe to the root of the German character."[217]

Calls for violence against Jews increased in the conservative reaction after 1848. Under a bloody red cross, rioters shouted "Murder the Jews!" "HEP HEP Destroy the Jews!" or "Freedom, one Religion, one God."[218] Volkstümler, a conservative writer, predicted that one day the Jews, along with their communist accomplices, would pay for their "sins and evil deeds by being beaten to death as before."[219] In 1848 and 1849, with antisemitism on the rise, violence spread throughout Germany and Austria against the Jews. In the next few decades, although individual personal relationships between Christians and Jews improved and some philosemitic literature appeared, antisemitism never lost its grip on German culture.

By 1871, after victory in the Franco-Prussian War united the Germanies, the new German Imperial Constitution granted civil liberties and religious freedom to all Germans, including Jews (only Switzerland, Norway, Spain, and Portugal still refused freedom of religion to Jews).[220] No sooner had the Jews achieved political freedom and equality, than a backlash occurred. Despite their contributions to almost all areas of German society, Jews were associated with the evils of modernization in a nation unprepared for change. The Jews paid for the prominence of a few Jewish families—the

Bleichröders, the richest German family of the time, the Rothschilds, Warburgs, and Oppenheims—in banking, business, higher education, medicine, law, journalism, and industry by widespread discrimination, especially in the officer corps, diplomatic service, and government.[221] The traditional principle that Jews were an inherently evil people manifested itself among German socialists, conservatives, and intellectuals.

German Socialists: Jews as Followers of Mammon

Germany, of all Christian countries, most closely identified the Jews with usury and economic exploitation. Centuries of Christian theologians from the Gospel writers, through Jerome and Chrysostom, Peter the Venerable and Bernard of Clairvaux, had identified the Jews as carnal and materialistic. Luther had vehemently attacked all Jews as usurers who should be hanged. Pejorative words such as *jüdeln* entered the German vocabulary to signify the way Jews talked, thought, and dealt with money. Christians involved in moneylending were called *Kristen-Juden*. Christian usury was termed *Judenspiess*, or Jews' spear. As during the Middle Ages, devil, heretic, usurer, and Jew were seen as one and the same.[222] Steven Aschheim writes of the transformation of "immutable, incorrigible . . . dehistoricized, hypostatized" Jewishness into a myth of all "the evils of bourgeois society."[223]

The socialists believed that Jews were the modern incarnation of the Antichrist and Mammon. In fact, the vast majority of Jews were not victimizers, but victims; not oppressors, but oppressed; not exploiters, but exploited; not rich, but poor.[224] The most significant talmudic interpretations of Torah taught charity, cooperation, and mutual respect, not usury. Jews did not initiate moneylending; they had been forced into it, and, as Abelard pointed out in the twelfth century, this profession made them seem more like pariahs than before. As the French-Jewish writer Bernard Lazare put it, the main cause of the Jews' predicament was Christian contempt.[225]

Bruno Bauer claimed that Jews invited Christians to hate and persecute them by their own cruelty and cowardice, their stubborn adherence to their religion "without love," their aloofness, their sterility, their self-conception as the chosen people, their hatred of other nations, and their attempts to dominate the world through money. He generally saw them becoming truly free only in a nonsectarian utopian state.[226] After 1848, Bauer joined the Prussian conservatives and dropped his demand that Christianity follow Judaism into oblivion.[227]

Although he was descended from rabbis, Marx was a self-hating Jew.[228] He repeatedly used Jew as an epithet and found Judaism "obnoxious" and "dirty."[229] Four years before his death in 1883, he mentioned that "Ramsgate is full of fleas and Jews."[230] Marx also hurled antisemitic insults at his Jewish opponents: Max Friedländer was "the cursed Jew of Vienna"; about Meier, Marx noted "that pig . . . is a Jew," Leo Fränkel was "a regular Jew."[231] Marx singled out his rival Ferdinand Lassalle for special attention. He was a "Jewish nigger" and "the most barbarous of all the Yids from Poland." (Lassalle was also a self-hating Jew: "I do not like the Jews at all, I even detest them in general. I see in them nothing but the very much degenerated sons of a great but vanished past." Engels called Lassalle a "real Jew from the Slav frontier." Engles felt that Frankel was a "real little Yid.")[232]

Baptized a Lutheran, along with the rest of his family, when he was six years old, Marx drew on Christian traditions to associate the Jews with the most evil abuses of money. He observed that whereas Bauer advocated an emancipation of Jews from materialistic Judaism, in reality "all of mankind had to be emancipated from Judaism."[233] He associated the Jews with the corrupt uses of money, an idea expressed by Paul, in the Gospels and the Church Fathers, through the modern period and was also put forward by Adolf Hitler: "It was the Jews, of course, who invented the economic system of constant fluctuation and expansion that we call capitalism. . . . It is an invention of genius, of the devil's own ingenuity. . . . But now we have challenged them, with a system of permanent revolution."[234] Marx accused the Jews of having destroyed human values, of having created capitalism, which in turn had "robbed the whole universe, the world of men, of nature, of their specific values."[235]

In *On the Jewish Question*, Marx held that the Jews had morally corrupted the initially superior Christianity. Through the Jew, "money has become a world power and the practical Jewish spirit has become the practical spirit of the Christian nations. . . . The Christians have become Jews."

The Conservative Antisemitic Campaign of the 1880s

The German conservatives believed that Germany was a "Christian state"—a conservative anti-Jewish and anti-liberal slogan among both Lutherans and Catholics[236]—a *Corpus Christianum*, to which the Jews did not belong. They charged that Jews embodied all that was "strange" and "inorganic"; they were religiously, culturally, and mentally alien; and they were interested only in business and liberalism.[237] During the financial disasters of the 1870s, the Conservative Party's journal, the *Kreuzzeitung* (also called the *Neue Preusische Zeitung*; the *Kreutzzeitung's* masthead contained the traditional

Prussian symbol, the iron cross) and its circle argued that the Jews and their allies controlled the German press, economy, and state.[238] Jews were to be granted hospitality and protection but not the right to positions of authority.[239] Jews were to be prohibited especially from the bureaucracy, judiciary, and education, for these institutions symbolized the Christian-German national spirit and must never fall into the hands of Jews.[240]

One of the most prominent conservatives of the period was Adolf Stöcker (1834–1909), court preacher of the Berlin Cathedral.[241] Patriotic and monarchist, Stöcker advocated a "Christian state" to keep the Jews in an inferior position. In the wake of economic hard times in the 1870s, this "second Luther" helped convince many of the German middle class, especially the middle and lower grades of professionals and small businessmen, the *Mittelstand*—Germans, neither capitalists nor wage earners, consisting of craftsmen, shopkeepers, farmers, civil servants, professionals (doctors, lawyers, teachers), students, and white-collar salaried employees—who had previously been unaffiliated with a political party, to support the conservative antisemitic parties.[242]

Stöcker's founding the Christian Social Workers' Party (later the Christian Socials) in 1878 provided him an influential role in German right-wing politics. In the election campaign of the same year he rehashed the patristic and medieval command that Jews must have no authority over Christians. "We firmly believe that no Jew can be [a] leader of Christian workers in either a religious or an economic capacity. The Christian Social Party inscribes Christianity on its banner."[243] Although Stöcker's Party lost influence by 1890, his conservative Lutheran anti-Jewishness had helped prepare the German masses to accept more readily the demagogic appeals of later agitators.[244] "[The Jews] foundered on Christ, lost their divine course, and surrendered their divine mission. . . . The Jews are and remain a people within a people, a state within a state. . . . Over against the German essence, they set their unbroken Semitism; against Christianity, their stubborn cult of the law or their enmity toward Christians."[245] His solution to the "Jewish problem" was to eliminate the Jews from all influential sections of German life. Only this "would restore Germany to blessedness, or the cancer from which we suffer will continue to eat away at us. Our future will then be imperiled, our German spirit will be Jewified."[246]

On the Jewish Day of Atonement (Yom Kippur), 26 September 1879, Wilhelm Marr founded the Antisemites' League, the first organization devoted exclusively to promoting political antisemitism. Apparently unaware of Christian racism and institutionalized Christian antisemitism, Marr distinguished between what he saw as religious Jew hatred and modern political-racial antisemitism; he hoped that his word "antisemitism" would move traditional anti-Jewishness into the secular realm by providing a veneer of respectability to the concept of Jew-hatred. Marr's organization reflected his secular racism, which existed alongside his religious antisemitism. On the one hand, like a racist, he argued that no Jew could be fully human.[247] Christians despised Jews because of their different "chemical" composition. "Jewishness itself . . . is disgusting."[248] On the other hand, although he defended Jews against the collective guilt of the crucifixion, Marr associated "Germanness" with Christianity and contrasted them both to Jewishness. Called "the new Luther" and defending Christian hostility to Jewish domination,[249] Marr believed that Germany was a Christian country, and his goal was "to free Christianity from the yoke of Judaism." The Antisemites' League used a German oak leaf and a Christian cross as its symbols. Hoping that Church and state could cooperate against the Jews, Marr's anti-Jewish political coalition's slogan was "Christianity, Kaiserism, Fatherland." In an 1891 article, Marr referred to his movement as composed of "the Christians and Aryans." ("Aryan" was invented by nineteenth-century philologists William Jones and Franz Bopp; Michaelis' disciple, Johann Gottfried Eichhorn, first labeled the Hebrew and Arab languages, "Semitic.") Marr's biographer, Moshe Zimmermann, has noted that "the real essence of Jew-hatred . . . remained anchored, more or less, in the Christian tradition, even when it moved, via the natural sciences, into racism."[250]

In 1881–82, Marr's Antisemites' League, university students, Stöcker's Christian Social Party, and most of the Catholic Center Party were instrumental in preparing, publishing, and presenting to the government the "Antisemites' Petition," also inspired by Wagner's 1880 Bayreuth Festival and Treitschke's writings. Containing nearly a quarter of a million signatures, this document sought to reverse the political and social gains Jews had made in Germany and to free Germany from "economic servitude" and "from a kind of alien domination" by limiting or preventing Jewish immigration and by excluding Jews "from all positions of authority," especially in the educational system. The Prussian minister of the interior refused to censure the petition. Encouraged, the petitioners in April 1882 presented it to Chancellor Otto von Bismarck, who used antisemitism for his own political purposes. (In 1847 at the Prussian Diet, Bismarck admitted his prejudice against Jews, which "I have sucked in with my mother's milk" and which he shared with most Germans. As chancellor, Bismarck also realized that his staunchest antisocialist allies, the conservatives and Catholic Center Party, would be pleased by his appearing to side with antisemitism.)[251] The petition stated, in part: "Wherever Christian and Jew enter social relations, we see the Jew as master and the native-born Christian population in a servile position. [The petition urges] that the Christian character of the primary school—even when attended by Jewish pupils—be strictly protected; that only Christian teachers be allowed . . . ."[252]

In 1881, the Prussian state prosecutor delivered a blow to the Berlin Jewish community when he held that the Talmud was not a religious work.[253] The prosecutor confirmed the medieval view that only the Jewish Bible as interpreted by Christian theologians deserved the status of Scripture. The Talmud was some peculiarly Jewish work that corrupted Jews.

German Intellectuals and the Antisemitic Campaign

Heinrich von Treitschke was an eminent conservative historian and editor of the influential *Preussische Jahrbücher*. The intellectual leader of German antisemitism in his day,[254] Treitschke—influenced by a Social Darwinism that gloried in a "pitiless racial struggle"—expressed what the German masses and many of the intellectuals of the time (and later) felt about the Jews: that they were obnoxious aliens whose presence was incompatible with Christian Germany. Associating assimilated and unassimilated Jews, he wrote, "There was nothing German about these people with their stinking caftans and their obligatory lovelocks, except their detestable speech."[255]

Treitschke was also a prophet of German-Christian nationalism, which, like that of other European nations, was often antisemitic.[256] He felt that the German-Jews were "a great danger, a serious sore spot of the new German national life." Emancipated Jews who demanded equality manifested "a dangerous spirit of arrogance." He attacked an authoritative history of the Jewish people written by the Jewish scholar Heinrich Graetz as "fanatical fury against the 'arch enemy' Christianity, [a] deadly hatred of the purest and most powerful exponents of German character, from Luther and Goethe and Fichte!" Treitschke concluded that "we Germans are, after all, a Christian nation . . . . We hear today the cry, as from one mouth, 'the Jews are our misfortune!' "[257] He was quoting Martin Luther.

Paul Bötticher, who later adopted his great aunt's name of de Lagarde, was a Christian critic of "dogmatic" Christianity whose works, like Treitschke's, were later used by the National-Socialists. Lagarde's biographer, Robert Lougee, points out that despite his attack on traditional Christianity, "none of his writings reflects doubt that the way of Christ is the way of God" and that his gravestone is carved with the epitaph, "The way of the cross is the road to salvation."[258] (Lagarde's father was a Lutheran minister, and the son had been destined for the church.) Lagarde's writing also expresses much of the traditional Christian assault on the Jewish people, whom he believed represented an alien culture in a Christian-German country, "a nation within a nation."[259] He considered the Jews as a people with different "views, customs, and ways of speech .... [I]n the midst of a Christian world Jews are Asiatic pagans."[260] Lagarde attacked Judaism as a desiccated, insipid legalism, which he called pharisaism, that was overattached to externals, and he believed that Jews committed ritual murder.[261]

Besides attacking Jews on traditional Christian grounds, Lagarde made the most virulent biological assault on the Jews in the nineteenth century.

> One would need a heart as hard as crocodile hide not to feel sorry for the poor exploited Germans and—which is identical—not to hate the Jews and despise those who—out of humanity!—defend these Jews or who are too cowardly to trample this usurious vermin to death. With trichinae and bacilli one does not negotiate, nor are trichinae and bacilli to be educated; they are exterminated as quickly and thoroughly as possible.[262]

At other times, Lagarde ridiculed biological racism, arguing instead that "Germanism is not a matter of the blood but of the spirit. . . . No idealistically inclined person can ever deny that spirit could and should conquer race."[263] Lagarde's violent anti-Jewishness was reason enough for the Nazis to distribute his works to German troops in the East in 1944.[264]

Like Lagarde, Konstantin Frantz was the son of a Lutheran minister. He asserted that the true German state was a Christian state;[265] its sovereignty rested on the recognition of Christ. In opposing Christ, Christianity, and the Christian state, Jews had rejected the entire cultural, political, moral, and religious authority of Christian Europe. They had become the Eternal Jew, impossible to integrate "into the Christian peoples and states. They can neither live nor die, and from this fate no earthly power can redeem them. . . . Through their own guilt, the Jews have excluded themselves from the community founded by Christianity, and no one can ever free them from that punishment. . . . The Jews in the Christian state are fundamentally rightless."[266] According to Frantz, Jews must be ghettoized and "pay protection money" and must never marry Christians.[267] Christianity had no room for the alien Jewish spirit.[268]

In the 1880s, another conservative historian, Hans Delbrück—assistant editor of the *Preussische Jahrbücher* and a specialist in the history of warfare—reflected common anti-Jewish attitudes when he opposed Jews gaining positions of power in Christian society "because they are Jews, and education, teaching, or the judiciary are regarded in our culture as the product of the Christian spirit . . . to which outsiders cannot contribute." He asserted that lacking an "authentic idea of Protestantism" Jews must not be allowed to insinuate themselves "into a select society to which [they] do not belong."[269]

In 1901 another leading Protestant conservative, Professor A. Suchsland, argued that the soul of the German people (*Volksseele*) and the Christian nature of Germany's institutions were in danger of being "de-Christianized" by Jews.[270] Denying the Jewish roots of Christianity, Adolf von Harnack, the outstanding patristic scholar of his time, argued that the Jewish Bible should be deleted from the Christian canon.[271] Another outstanding German theologian and New Testament scholar, Rudolf Bultmann, noted that Jesus' Gospel was in fundamental "opposition to the morality and piety of Judaism."[272]

There existed, among Jews and Christians, opposition to antisemitism. A leader was the liberal historian Theodor Mommsen. The major Christian intellectual challenging Treitschke's antisemitism,[273] Mommsen attacked the antisemitic campaign and called Treitschke's "suicidal agitation of national sentiment" one of "the most silly perversions" and "a national calamity."[274] But the liberal Mommsen rejected the Jews' desire to maintain their cultural and religious identity. "We cannot . . . protect the Jews from the estrangement and inequality with which the German-Christian still tends to treat them. . . . *Christianity* . . . is the only word which still defines the entire international civilization of our day . . . . He whose conscience . . . does not permit him to renounce his Judaism and accept Christianity . . . should be prepared to bear the consequences."[275] Mommsen also asserted that the Jews comprised "an element of decomposition" of individual Germanic tribes, which led to a common German nationality; the development of Germany required the addition of "a few percent Israel" as a catalyst. Mommsen later explained that he had not intended the word "decomposition" to demean the Jews but to compliment them.[276] Nevertheless, his phrase was widely misunderstood at the time as an insult to the Jews (by Theodor Fritsch and Houston Stewart Chamberlain) and was later quoted by Hitler in several speeches—on 12 April 1922, 27 January 1932, 2 March, 20 April, and 1 September 1933, 8 October 1935, 1 January 1938[277] and Hitler mentioned it in his table talk: "This destructive role of the Jews [as] the ferment that causes people to decay . . . provoke[s] the defensive reaction of the attacked organism."[278]

Catholic Attacks on the Talmud

Throughout the century, German Catholic theologians had repeated the medieval charges against "talmudic anti-Christianity" and argued that Christian culture would be demolished if Jews were to be emancipated and granted equal rights.[279] The Jesuit journal, *Civiltà Cattolica*, began an international anti-Jewish campaign during this period. In 1882, the First International Anti-Jewish Congress was held at Dresden with delegates from all over Germany, Austria, and Hungary. Citing the Church Fathers, Catholic spokesmen contended that, from the beginning, Jews were immoral and subversive betrayers and corrupters of Christian nations, including Germany.[280]

In the same year, August Rohling, priest and professor of Catholic theology and Hebrew Antiquities at the prestigious Charles University of Prague, testified as an expert witness at the Tísza-Eszlár murder trial. Fifteen members of the Jewish community of this Hungarian town were accused of murdering a Christian child for her blood—the blood-libel defamation was widely believed in the Hapsburg monarchy. Between 1867 and 1914, there were a dozen ritual-murder trials in Austria-Hungary, where Jews were "caught in a crossfire of ethnic and social conflicts," Magyars against Slavs, Slavs against Germans, and everyone against the Jews.[281] The prosecution called Rohling, whom Richard Wagner admired,[282] as an expert witness because in 1871 his *Der Talmudjude*—based on Eisenmenger's *Entdektes Judenthum*—had alleged that the Talmud, in contrast to the Holy Bible, was a Jewish collection of immoral and irrational superstitions and that the Talmud required Jews to hate Christians, to practice ritual murder, and to seek domination of the world. Although the accused Jews were all acquitted, Rohling testified that "the religion of the Jews requires them to despoil and destroy Christianity in every way possible [and that] the shedding of a Christian virgin's blood is for the Jews an extraordinarily holy event."[283] Rohling's book continued to have a large readership, because so many Christians wanted to believe in the "factuality" of the ritual-murder defamation.[284] Once Rohling's themes were picked up by the Jesuit journal *Civiltà Cattolica* in its late nineteenth-century antisemitic campaign, it was as if these Judeophobic ideas were awarded a papal imprimatur.[285]

The case drew national and international attention when the local Catholic priest reported it to the press and appealed to the leading Hungarian antisemitic politician, Gyözö Istóczy, to help prevent the case from being "suppressed" by the Jews.[286] The trial was followed by pogroms in Hungary, where Istóczy's antisemitic party won more than 20 seats in the 1884 parliamentary elections.[287]

Wagner: Renegade Christian Antisemite

Richard Wagner was the most important German cultural figure of the nineteenth century. Many writers have seen Wagner as essentially a racist,[288] but the evidence shows that Wagner subscribed to Count Joseph Arthur de Gobineau's racial ideas for only a few years and that Wagner often attacked biological racism.[289] In his most popular anti-Jewish work, "Judaism in Music," Wagner did not use the word *Rasse*, "race." At other times, he employed words such as *Stammes* (stock, tribes, families, clans), meaning ethnic, religious, or cultural groups, that have been translated into English as "race" but lacked any racist meanings.[290] He wrote that "the art-work of the future must embrace the spirit of a free mankind, . . . its racial imprint must be no more than an embellishment, the individual charm of manifold diversity and not a cramping barrier."[291] In another work, he contrasted the "purely human" with a defective "race-religion"—the ancient Indian Aryan religion.[292] Writing in 1881 to the Jewish theatrical director Angelo Neumann—formerly Wagner's favorite tenor—Wagner claimed to have "nothing to do with the current [racist] antisemitic movement."[293] Wagner's wife, Cosima,[294] recorded that "over dinner, he regularly exploded in favor of the Christian point of view as opposed to the racial one."[295] Rather than assuming that blood determined race, Wagner believed that the blood of Christ alone provided the possibility of salvation. "One thing is certain," he said, "races are done for, and all that can now make an impact is . . . the blood of Christ."[296] Wagner also wrote about "the blood of the Savior, the issue from his head, his wounds upon the cross—who impiously would ask its race, if white or other?"[297]

Wagner was raised as a Christian, and his emotions were conditioned by the same Christian images, rhetoric, and myths as his audience. At one point Wagner said that although the Christian spirit was corrupted by the Church, "It was the spirit of Christianity that rewoke to life the soul of music."[298] Wagner's experience of church rituals, like the young Hitler's, was intoxicating.[299] At his first communion at 13, he remembered "The shudder of emotion when the bread and wine were offered and taken . . . ." And Wagner admitted in his autobiography that he longed to assume Jesus' place on the cross on the altar of his boyhood church.[300]

Wagner's perception of the Jews as evil beings was based on traditional and contemporary Christian thought and dominated much of his mental and emotional life. Like Hitler, Wagner was bedeviled by the possibility that he was part Jewish. Wagner seemed to believe that his biological father was Jewish, and from his childhood, as he wrote, "I felt a long-repressed hatred for this Jewry, and this hatred is as necessary to my nature as gall is to the blood."[301] When he described "our natural repugnance against the Jewish nature," like so many other Christians he was referring to what he learned as a child being raised in a Christian family and environment. He wrote about

> that involuntary feeling of ours which utters itself as an instinctive repugnance against the Jew's prime essence. . . . [M]ay we even hope to rout the

> demon from the field, whereon he has only been able to maintain his stand beneath the shelter of a twilight darkness . . . to make him look less loathsome.[302]

Following in the intellectual footsteps of earlier German thinkers such as Herder, Kant, and Hegel, Wagner's continually repeated goal was to reform Christianity and Christianize the Jews. An anti-ecclesiastical Christian, not a faithful churchgoer, but a believer in his own kind of Christianity,[303] Wagner thought that Jews could overcome their Jewish spirit through conversion to his kind of Christianity.[304] Wagner sometimes saw Jewish redemption as occurring in two stages: first, the Jew had to renounce his Jewishness; second, the Jew had to become a Wagnerian kind of Christian.[305] Wagner tried—and failed—to have the great conductor, Hermann Levi, son of a rabbi,[306] baptized so that as a Christian he could conduct the premiere of *Parsifal*, Wagner's most Christian opera—Wagner's own analysis of the opera's themes listed the Christian values of "Love—Faith—Hope" and in a letter to King Ludwig II of Bavaria, Wagner wrote that *Parsifal* was "this most Christian of works."[307] "I cannot allow [Levi] to conduct *Parsifal* unbaptized," he told Cosima.[308]

(On 11 October 1881, King Ludwig wrote to Wagner concerning the choice of Hermann Levi as conductor: "[There is] no distinction between Christian and Jew. There is nothing so nauseous, so unedifying, as disputes of this sort: at bottom all men are brothers, whatever their confessional differences." Ludwig's highly moral beliefs were extraordinary for the time. Ironically, he was declared hopelessly insane in 1886.)[309]

In several essays, including the one Wagner considered his most important, "Herodom and Christendom," he explained that the Jews represented the multifaceted power of evil: Jews were "the plastic demon" responsible for the decadence of all human society.[310] Wagner felt that not only had the Jewish spirit distorted the life-giving principles of compassion and love contained in authentic Christianity, but it had also distorted the church, troubled his nation, corrupted Western civilization, and intended to dominate the world.[311]

The cultural trait that Wagner most hated was materialism—this was his public position, but privately, despite his incessant preaching about great art, he was passionately concerned with money—and he accused the Jews of being the corruptive force behind it. Jewish money, "the guileless-looking scrap of paper[,] is slimy with the blood of countless generations."[312] For Wagner, the Jew was a "reckoning beast of prey," whom Christian-Germans have allowed to dominate the financial aspects of society.[313] Those, like the Jews, who seek "power, dominion—above all, the protection of property" commit sin and murder life itself.[314] Wagner put the matter less ceremoniously in a letter written during the 1840s, when he called moneylending, "damned Jewish slime."[315] Wagner traced Jewish materialism to the Torah, which he believed convinced the Jews to reject Jesus' antimaterialistic and life-giving message of love and to murder him. (In *Jesus of Nazareth*, Wagner has the Jews cheer when Pilate turns Jesus over to the soldiers for crucifixion.)[316] Like Hitler, Wagner doubted that Jesus "was of Jewish extraction."[317] "Jesus frees our human nature," wrote Wagner, "when he abrogates the [Jewish] law [and] proclaims the divine law of love . . . ."[318] Wagner wrote to Constantin Frantz that "If the common people were made to forget about God in the 'burning bush' and shown instead only the 'sacred head sore wounded,' they would understand what Christianity is all about. . . . [T]he way from man to Him is *compassion*, and its everlasting name is *Jesus*."[319]

Hitler claimed inspiration from Wagner, but we have no real proof of it. Whether Wagner would have objected to Hitler's radical answers to the Jewish problem is unknowable. Wagner himself was not clear about what to do with the Jews. Like Schopenhauer, Wagner believed that the only good Jew was a convert to a dejudaized Christianity.[320] In his 1869 Appendix to "Judaism in Music," Wagner wondered whether the Jewish corruption of Gentile culture could be blocked "by a violent ejection of the destructive foreign element."[321] He may have entertained the same solution to the political chaos in Germany, which he blamed on the Jewish spirit. (He considered democracy an unGerman mishmash of French and Jewish ideas.)[322] "Only when . . . party strife no more can find a where or when to lurk among us, will there also be no longer—any Jews."[323] Wagner claimed that he could not decide about what was to be done with the many stubborn Jews who were not only incapable of artistic endeavor, but who also refused to be baptized into Wagnerian-style Christianity.[324] He was unsure about a violent solution, since it "would require forces with whose existence I am unacquainted."[325]

Wagner's letter to King Ludwig II of Bavaria concerning Jewish musicians indicated that Wagner, like Hitler, saw himself as the artist standing as Germany's last bastion against the power of the "evil" Jews.

> I consider the Jewish race the born enemy of pure humanity and all that is noble in man: there is no doubt that we Germans especially will be destroyed by them, and I may well be the last remaining German who, as an artist, has known how to hold his ground in the face of a Judaism which is now all-powerful.[326]

Both *Mein Kampf* and *Secret Conversations* are packed with references to this attitude of Hitler's. Ernest Newman calls Wagner's letter a "charming specimen of Hitlerism *avant la lettre*."[327]

We do know how Wagner reacted in the face of deadly violence perpetrated against Jews during his lifetime. According to Cosima's diaries, Wagner approved the massive Russian pogroms of 1881, welcoming the violence as "an expression of the strength of the people."[328] In the same year, about 800 Jews and Christians died in a fire at Vienna's Ringtheater, where the audience was predominantly Jewish.[329] They were attending a performance of the French-Jewish composer Jacques Offenbach, and Wagner observed that "the most useless people frequented such an opera house."[330] A few days later, he tried to justify his response:

> It sounds hard and is almost unnatural, but people are too wicked for one to be much affected when they perish in masses . . . . When such-and-such a number of members of this [Jewish and Christian admirers] community die while watching an Offenbach operetta, an activity which contains no trace of moral superiority, that leaves me quite indifferent.[331]

Two days later, according to Cosima, Wagner made it quite clear that he had been thinking about the Jews when he had offered these cold and cruel comments about the Vienna fire, for he "made a drastic joke to the effect that all Jews should be burned at a performance of [Lessing's] *Nathan* [*the Wise*]."[332]

Austrian Political Antisemitism in the Last Decade of the Century

Jews had lived in Vienna since 1200, but were expelled in 1421, 1670, and 1938. Austria's antisemitism was deeper and more pervasive than that of Germany, in part because many Austrian Jews were more religious than their German counterparts. Although between 1889 and 1938 Jews comprised only ten percent of Vienna's population, they supplied a high proportion of its creative geniuses and helped give the city its sparkling reputation.[333]

The leading Austrian antisemitic politician at the end of the nineteenth century was Karl Lueger. Mayor of Vienna, representative to the Austrian Parliament, and head of the Christian-Social Party, Lueger drew many of his ideas on Jews from the works of Abraham a Sancta Clara, the most important Catholic preacher of the seventeenth century and a virulent antisemite. Trained by the Jesuits, Sancta Clara felt that the Jews were one of Christendom's prime enemies and classified Jews along with gravediggers and witches as causes of the plague. He accused the Jews of desecrating the

Host and murdering Christian children for the sake of the devil. Like Luther, he ranted that "besides the devil, the Jews are the worst enemy of mankind. . . . Jewish beliefs are such that all Jews should to be hanged and burned."[334]

Lueger and the priest Ignaz Seipel, future chancellor of Austria, were both disciples of Karl Vogelsang, the founder of social Catholicism. Vogelsang was not a racist, but a traditional antisemite who argued that Jews were opposed to justice and Christian community.[335] Lueger headed the Christian-Social Party, familiarly known as the Antisemitic Party. It was a Catholic political party supported by middle-class businessmen, the faithful, and the priesthood, and ultimately endorsed by the papacy. The Hapsburg regime and the Austrian Catholic Church hierarchy initially condemned Lueger's party and his election as Vienna's mayor.[336] But in 1894 and 1895, Pope Leo XIII blessed and commended the Christian-Social Party and its newspaper, the *Reichspost*. To obtain the pope's support, Lueger had assured him that his party and his own antisemitism were traditional, not "racial." (Lueger's party republished several antisemitic articles (which Hitler may have read) taken from the Jesuit publication *Civiltà Cattolica*.) The pope, whose desk was later adorned with Lueger's picture, replied, "The leader of the Christian Social Union may know that he has in his pope a true friend who blesses him and treasures the Christian Social effort." Pope Leo's secretary of state, Cardinal Rampolla, rejoiced at the party's overwhelming victory in Vienna's municipal elections in the mid-1890s, "You see, we have triumphed."[337] With this kind of endorsement, Hapsburg emperor Franz Joseph accepted Lueger as Vienna's mayor in 1897, where he and his Antisemitic Party ruled until his death in 1910.

Lueger openly admitted the opportunistic element in his antisemitism. In reply to criticism that he accepted too many Jewish dinner invitations, Lueger replied that "I myself decide who is a Jew."[338] But his actions and statements more often reflected traditional Christian antisemitism, which help explain his great appeal to all levels of Viennese society. In an 1890 speech before the Austrian parliament, the Reichsrat, Lueger pictured the Jews as "unbelievably fanatical [in their] hatred and their insatiable love of revenge. . . . What are wolves, lions, panthers, leopards, tigers, and men in comparison with these beasts of prey in human form? . . . We do not shout 'HEP, HEP,' but we object to Christians being oppressed . . . ."[339] He remarked that "only Christian antisemitism is national and effective; it measures up to every standard of culture and humanity and proceeds naturally from the eighteen hundred years of Christian life and teaching . . . ."[340] In 1897, Lueger wrote in his party's newspaper that Christian antisemitism "fights Jewish treachery against Christian teaching and culture, against a Christian society and state.[It] sees to it that Christian people remain masters in their own home."[341] In the same year, he supported a bill in the Reichsrat designed to end Jewish immigration into the empire. The bill described the Jews as foreigners and "enemies of Christian culture and of nations of Aryan descent."[342]

The antisemitic political parties' decline in voting strength toward the end of the nineteenth century was only temporary.[343] Anti-Jewish arguments and violence against Jews, especially among the youth, became a respectable part of public opinion in part because of widespread anti-Jewish propaganda. German and Austrian political antisemitism would intensify once these nations experienced the disastrous political, economic, and social crises in the twentieth century.[344]

The German and Austrian Jews were emancipated, but Jewishness was not. Christianity was almost universally recognized as superior, Judaism as inferior.[345] Nor were German-Christians "emancipated" from their basic religious antagonisms toward Jews. The culture and thought of Christian Europe still treated Jews with varying degrees of alienation and antipathy.[346] Christian-Germans still distanced themselves from Jews socially, politically, and legally. They still associated Jews with the immoral use of money and with a stubborn attachment to a religion that lacked divine sanction.[347] Average Germans,[348] who bought Streicher's Nazi newspaper, *Der Stürmer*, in post–World War I Germany may have felt that the paper went a bit too far in its anti-Jewish onslaughts, but they felt very comfortable with the basic anti-Jewish values it proclaimed.[349]

Chapter 5

Christian Antisemitism, the German People, and Adolf Hitler

In boundless love, as a Christian and a human being, I read through the passage that tells us how the Lord arose at last in His might and seized the scourge to drive out of the Temple the brood of vipers and adders. How terrific was His fight against the Jewish poison. Today, after two thousand years, with deepest emotion I realize more profoundly than ever before the fact that it was for this that He had to shed His blood upon the Cross.

—*Adolf Hitler*

Several early-twentieth-century historical events made the "Final Solution of the Jewish Problem" possible in Germany. In other nations, including the Allied nations, there was also a high level of Christian antisemitism, but it was held in check by countervailing traditions of democracy, civil liberties, justice, and so forth. In Germany, these opposing trends were not strong. In the context of Christian antisemitism and contemporary crisis, Hitler and the Nazis and their collaborators and even their German opponents scapegoated the Jews. It was widely accepted that the Jews stood behind Germany's troubles—Marxism was seen as a Jewish political movement; homosexuality and mental and physical deformities as Jewish diseases; Jehovah's Witnesses as in league with the Jews; Gypsies as satanic mongrels intermarried with the Jews; Soviet Russia as the political arm of the Jews; criminals as following Jewish behavior patterns.

Hitler and the Nazis made sympathetic connections with almost every layer of German society—except the far Left—and with millions not within Germany. This was made possible by a shared Christian antisemitism. The German political, social, and religious elites, the farmers, the middle classes made the National-Socialist German Workers' Party modern Germany's most popular political party. From the beginning of the post–First World War period—years before Hitler came to power, years before the outbreak of the Second World War, years before the death factories of the Third Reich—German awareness of the violence of Nazi language and of the Nazi promises to save Germany by destroying the Jews did nothing to diminish the enthusiasm of those Germans who supported the regime and therefore the Final Solution of the Jewish Problem.

In the first half of the twentieth century, the German people suffered three decades of repeated trauma. Even though, compared to the other major nations engaged in the First World War, Germany was responsible for a plurality of guilt, and granted that German national territory never suffered combat during 1914–18, nevertheless the events between and including two World Wars were disastrous: loss of the war after German leadership had repeatedly promised victory, the Versailles peace treaty that was neither harsh enough nor conciliatory enough (as John Maynard Keynes pointed out, it "fell between two stools"), the coerced establishment of a Weimar Republic, history's worst inflation in 1923, an economic Depression, political and social chaos, the establishment of the Hitlerian dictatorship, the experience of the Second World War. One cannot fail to connect these events with the Holocaust. This hectic history, along with psychosocial disasters such as the fear of annihilation, fantasies of betrayal and divine election, revelation of conspiracies, and redemptive political forces caused the kind of delusional Judeophobic thinking that resulted in the Holocaust.[1] These same factors were alive and well in the Middle Ages, consisting of delusional beliefs about Jews, the fictionalization, misperceptions, and mythologizing of Jews: the alleged inherent Jewish stubbornness, wanderings, satanic connections, ritual child murder, vampirism, desecrations of Christian sacraments, conspiracy against Christendom.[2] As shown in the beginning chapters of this book, even earlier, in the first centuries of the Christian era, the Church Fathers accused Jews of deicide, devilishness, betrayal of God for money, and sacramental defilements. Various ideologies played their part in twentieth-century German antisemitism: racism, Social Darwinism, geopolitics, Volkish nationalism, antimodernism, and anti-Bolshevism were important. But it was the long-term and the short-term influence of Christian theological antisemitism and Christian racist antisemitism that provided the most important roots of the Holocaust.

Christianity and the Third Reich

The Christian Churches obviously did not perpetrate the Final Solution. Indeed, Christian ethical principles are so antithetical to the genocidal morality of Nazi Germany that no connection seems evident between any Christian precepts and the Final Solution. Most scholars recognize, however, an important duality in Christian attitudes toward the Jews that contributed significantly to the carrying out of the "Final Solution." On the one hand, a miniscule minority of authentic Christians,[3] acting on Jesus' moral teachings, helped the Jews, often at great risk to themselves. On the other hand, a much larger minority of Christians attempted to kill all the Jews of Europe. Most other Christians either actively collaborated in this murderous endeavor or tacitly permitted it to happen.[4] Their behavior reflected Christian anti-Jewish principles elaborated over nearly two millennia.

The Churches and their theologians had formulated compelling religious, social, and moral *idées forces* that provided a conceptual framework for the Nazi Holocaust. The National-Socialist conception of the Jew as less than human, or inhuman, was based on the Christian conception of the Jew as traitor, murderer, plague, pollution, filth, devil, and insect, which "prepared both killers and victims for the Jews' literal destruction."[5] Jewish deaths at the hands of Christians before 1933 amounted to millions. In the Christian mind, as John Bossy has put it, "the Jews were the original enemies of Christ, who had procured his crucifixion and death and had taken his blood upon their heads and upon those of their children. Nothing was easier for the average Christian to understand than that this was a crime that cried out for vengeance . . . ."[6] And over the nearly two millennia of the Christian era, vengeance has been taken by means of Church policies of degradation and only half-hearted protection during pogroms and mass murder of Jews from the Middle Ages on. As Léon Poliakov has put it, Jews have provided "an indispensable reference group, enabling Christians to know themselves as Christians and to incarnate good by contrast with [Jewish] evil."[7] Richard Rubenstein has observed, "it may be impossible for [most] Christians to remain Christians without regarding Jews in mythic, magic, and theological categories."[8] Luther wrote that Jews are a "base, whoring people, that is, no people of God, and their boast of lineage, circumcision, and law must be accounted as filth."[9] History awaited the right leader, movement, crisis, and context to actualize Christianity's antisemitic ideology into reality.

Without denying the Third Reich's horrendous level of coercive terror, its promise of great economic gain, and the secular-ideological attractions that led many decent people to collaborate with the regime, there could have been no National-Socialist Third Reich without a preexisting consensus on many issues.[10] As Peter Fritzsche has written: "Germans became Nazis because they wanted to become Nazis and because the Nazis spoke so well to their interests and inclinations."[11] Nationalism and economic reform may have been more important than antisemitism in attracting large numbers of Germans to vote for Hitler—after all, most German political parties carried antisemitism as a plank in their platform. Nevertheless, Christian theological and racist antisemitism prepared, conditioned, and encouraged a larger number of Germans—led in their religious prejudice by Catholic and Protestant Church authorities—to collaborate with Hitler's regime and to accept Hitler's Final Solution to the Jewish Problem, and a smaller number actively to collaborate. The Gestapo itself would have been ineffective if the general public had not voluntarily collaborated with government officials.[12] A 1947 report from American occupation authorities in Germany indicated that "those most infected with antisemitism were those who practiced the two Christian Confessions."[13] Fifteen years after the Holocaust chronologically ended, when scores of Germans were interviewed at the time of the Eichmann trial, from those more than 40 years old "as often as not we heard the frankly stated opinion that the Jews had brought the Final Solution on themselves. How so? 'Because they crucified our Lord Jesus Christ!' "[14] Typical was the National-Socialist Women's Organization, *Nationalsozialistische Frauenschaft*. One of its leaders, Hildegard Passow, noted that "at no time does the Lord God require of us charitable conciliation with the Jews, the moral enemy of the Aryan character. Christ himself called the Jews 'the sons of the Devil, a brood of snakes,' and drove the dealers and moneychangers from the house of God with a whip."[15] Twenty years after the Holocaust, Jewish historian Richard Rubenstein interviewed German Protestant hero and Christian resistance leader Heinrich Grüber, who had risked his life to help Jews and ended up incarcerated at Dachau. Grüber shocked Rubenstein when Grüber told him that he still believed, 20 years after the Holocaust, that Jews were murdered as part of God's plan and that although what Hitler had done was immoral, he was acting essentially as God's instrument. For God's continuing anger at the Jews required that they be punished, and Hitler was simply one of the "rods of God's anger." Grüber was saying that "the Christ-killers got what was coming to them."[16]

In his book on the German Churches and the Holocaust, Robert Eriksen argued that Hitler "was not a Christian, but an advocate of his own worldview of German racial destiny."[17] What makes a Christian? If someone other than Hitler were baptized, educated, and trained as a Christian, often spoke privately and publicly as a Christian, admired Jesus, quoted the

Bible, and so forth, would he not be considered a Christian? If that anonymous someone also made monstrously evil moral choices, would not most say, "He's a bad Christian"? Why would one describe him otherwise? When Moslems hijacked airplanes and flew them into the twin towers in New York City, one could argue they were bad Moslems but still, they were Moslems. Like so many other Christians, including Goebbels and Goering, Hitler had some cockeyed racial ideas; but all three were Christians.

Germany was a nation of Christians who took their religion very seriously.[18] In 1939, only 1.5 percent of Germans considered themselves unbelievers. Ninety-four percent of the population still belonged to the Protestant or Catholic Churches.[19] (Of those belonging to churches, 40 percent identified themselves as Catholic, 54 percent as Protestant. Once the Anschluß of Austria took place in March 1938, the proportion was roughly 50:50.) As in almost all of Europe, virtually all citizens identified themselves with some Christian faith or group. The vast majority of Germans, even those not formally affiliated with a church, were products of a Christian culture, family, and training. In nearly every home, Germans hung crosses on the walls; German families baptized, instructed, and confirmed their children within the church; Germans were married and buried in the church; Germans celebrated Christian holidays. Many important Nazis emphasized that they were "good Christians," some having served as ministers, church officials, and theology students.[20] From 1933 on, Catholic schoolchildren in Catholic schools in Germany were taught in their religious instruction about "the close affinity between Cross and Swastika."[21] (The *Hakenkreuz*, or swastika, was a variation of the Christian cross and was personally designed by Hitler.)[22] Nazi SA (Sturmabteilung) members, often with the approval of their Protestant ministers, marched to worship in churches draped with Swastikas.[23] Even though the Nazi government arrested Martin Niemoeller, the Nazi authorities permitted him to attend Protestant church services when he was incarcerated in Moabit prison. At Christmas time, the prison hall was full of Christmas trees.[24] At the concentration camp at Westerbork, Holland, the SS (Schutzstaffel) guards and their girlfriends celebrated the birth of Jesus with a Christmas party.[25] From a surrounded Stalingrad about to surrender in December 1942, the soldiers broadcast their rendition of Christmas carols back to Germany. There was a Catholic Chapel at Dachau concentration camp.[26] Even the most anticlerical of Nazis, Martin Bormann, defended the right of those who had left the church to have church bells rung at their funerals.[27] Although Bormann was anticlerical as Himmler was anticlerical, their hostility was not ideological, it was not based on antipathy to the Christian religion but on the Church's seeking after secular power. However Bormann turned in his anticlericalism, other leading Nazis—including Hitler himself and Goebbels—hindered his way.[28]

Even though most Germans put the Final Solution out of their minds, were concerned with more pressing issues, rationalized away what was happening, or feared challenging the authorities, nevertheless, to succeed in its objective to murder the Jews, the Hitler regime needed the passive sanction of all these "decent" Church-going Germans.[29] The acquiescence of the Churches was absolutely vital to enable the Reich to sustain the morale of the German Armed Forces and carry out its other policies, including the Holocaust. Although he planned to bring Christian institutions completely under the control of the Reich and in private called Christianity "an invention of sick brains; one could imagine nothing more senseless, nor any more indecent way of turning the idea of the Godhead into a mockery,"[30] Hitler at other times attributed this radical solution to his youth and argued instead that the churches should wither away of their own accord. "The best thing is to let Christianity die a natural death, . . . the rotten branch falls of itself." In this regard, he spoke of continuing state support (a grant of 50 million marks) for the Catholic Church after the war, at the same time making the recruiting of priests difficult.[31] Hitler was ambivalent about Christianity as an institution. At the least, he recognized that he had to speak publicly in favor of Christianity to indulge the German people's need to believe in their traditional religion.

The Christian Leadership

Alan Davies has written, "without the Church, Hitler would not have been possible."[32] Especially during the first years of the regime, Christian beliefs were not seen as inconsistent with Nazi ideology nor with state antisemitism.[33] The Protestant and ultimately the Catholic leadership, those who had the best chance to oppose Hitler, instead endorsed the regime wholeheartedly. Just as apples do not fall far from the tree, so the Protestant and Catholic parishioners followed their pastors' and priests' lead. The majority of both Protestants and Catholics endorsed the Nazi regime.[34] Conditioned by a bimillennial theological and racist Christian tradition hostile to the Jews, German lay Christians and Churches supported the Third Reich and its antisemitic policies. In the March election, many who voted for other parties still sympathized with the Nazis' policies and attitudes toward Jews.[35] Christians accommodated Nazism, shared its values, "fervently affirmed it."[36] "Through the support for Nazi policies articulated by many religious leaders, ordinary Germans were reassured that those policies did not violate the tenets of Christian faith and morality."[37]

The German-Christians as well as the "opposition" Confessing Church, *Bekennende Kirche*, led by Dietrich Bonhoeffer and Martin Niemoeller, sympathized with the Third Reich's Jewish policy. Not only that, some members of the Confessing Church were also members of the Nazi SA.[38] Almost all of the Catholic bishops supported the regime—Berlin bishop Konrad von Preysing being a major exception. Popes Pius XI and XII were both Germanophiles.[39] Pius XII never clearly exposed what he knew of the Holocaust, never clearly expressed sympathy with the Jews, never reminded Catholics of their moral duty as Christians with regard to the mass murder of civilians. In a speech of 20 July 1937, while still Pius XI's secretary of state, Pacelli unwittingly condemned himself: "It is at times of crisis that one can judge the hearts and characters of men."[40]

Perhaps the attitude of the majority of Germans was like that of the philosopher Martin Heidegger. He did not sympathize with and then join the Nazi Party only because of his antisemitism, but neither did the Nazis' anti-Jewish brutality deter him from supporting them. He accepted it all.[41] Likewise, the eminent Protestant theologians Gerhard Kittel, Paul Althaus, and Emanuel Hirsch were intelligent and respectable men as well as supporters of Adolf Hitler. (Long after the war and Holocaust had ended, Althaus' son asked him how he could have considered Hitler "a gift and miracle of God" and how he could have accepted the mass murder of millions of Jews? "You have not experienced the Jews," Althaus told his son.)[42] For them, and for the Nazis, and for most Germans, God's curse on the Jews was clear, and they would not oppose the Nazi government's policy toward the Jews. Historian Robert Erickson concluded that these theologians felt themselves "on the same side [as Hitler] of the *Weltanschauungskampf* . . . these three theologians saw themselves and were seen by others as genuine Christians acting upon genuine Christian impulses."[43] Their attitudes toward the Jews were nearly inseparable from those of Dietrich Bonhöffer and Martin Niemöller in the first few years of the Nazi regime.

On 1 April 1933, the German-government sponsored a boycott of Jewish shops and businesses and barred Jewish professionals from entering their offices. Dietrich Bonhöffer, the eminent Lutheran minister and theologian, a man who later helped German-Jews and himself was executed by the Nazis, defended the Reich's anti-Jewish action: "The state's measures against the Jewish people are connected . . . in a very special way with the Church. In the Church of Christ, we have never lost sight of the idea that the 'Chosen People,' that nailed the Savior of the world to the cross, must bear the curse of its action through a long history of suffering."[44] Even though he later struggled to help Jews and noted that Christ was a Jew, he still argued as late as 1941 that "The Jew keeps open the question of Christ. He is the sign of the free mercy-choice and of the repudiating wrath of God."[45]

Niemöller—like Bonhöffer, Friedrich Greunagel, Christian Stoll, Walter Kunneth, Rudolf Homann, Adolf Schlatter, Eduard Putz, and Hugo Fleming—took the same position on the Jewish "problem."[46] In a sermon of August 1935—three weeks before the first anti-Jewish Nuremberg Decrees were issued—Niemoeller drew a half-dozen parallels between the Nazis and their German supporters, and the Jews.[47] Niemöller specified that Jews stood as the standard by which he judged (and God would judge) the Nazis and their followers for their actual and potential shortcomings. That this anti-Jewish sermon, among others, was chosen by him and his publishers to be printed in England and the United States makes a statement about their notion of the international acceptability of the antisemitic ideas expressed therein.[48] In an adulatory Foreword to the American edition of the sermons, James Moffatt, a distinguished professor of Church History at Union Theological Seminary, wrote that the sermons were a rallying call and "the commanding word of God . . . . What [Niemöller] has to say is not new, but it is always needed." Indeed! The August sermon is replete with Christian myths about Jews, with Niemöller closely following the traditional path laid out by the Churches for nearly two millennia.

Like the Church Fathers, medieval theologians, Martin Luther, and the Nazis, Niemöller presented the Jews as the paradigmatic evildoers in Christendom. Pastor Niemöller sermonized that Jewish history was "dark and sinister" and that the Jewish people could neither live nor die because it was "under a curse, [that is,] "the 'eternal Jew' conjure[s] up the picture of a restless wanderer who has no home and who cannot find peace." In other words, the Jews are "a highly gifted people which produces idea after idea for the benefit of the world, but whatever it takes becomes poisoned, and all that it ever reaps is contempt and hatred because [always] the world notices the deception and avenges itself in its own way."[49] Niemöller's observations here seemed terribly close to those of Hitler's propaganda minister Joseph Goebbels, who in a speech of May 1943 asked:

> What will be the solution of the Jewish question? . . . It is curious to note that the countries where public opinion is rising in favor of the Jews refuse to accept them from us. They call them the pioneers of a new civilizaton, geniuses of philosophy and artistic creation, but when anybody wants them to accept these geniuses, they close their borders; "No, no. We don't want them!" It seems to me to be the only case in the world history where people have refused to accept geniuses.[50]

Martin Niemöller had been in a unique position to do good when it would really have counted. He was, after all, the most important leader of the most significant dissenting institution in Germany, the Confessing Church. That is why his distorted vision of the Jews was so damaging. In the context of the early years of the 1933–45 Holocaust, it was doubly destructive for him publicly to state the traditional antisemitic mythology about Jews as well as to compare the Nazis and the German-Christians with the Jewish people, making the Jews the prime criterion of evil. The net effect of his argument was to weaken the already fragile German resistance by making obvious the essential agreement, aside from the issue of Christians of Jewish background, between the Protestant leadership and the Nazis on pre-1941 Jewish policy. Their mutuality on this issue, in turn, helped make prevention of mass murder of Jews during the war impossible.[51] After his failed attempt to volunteer for the German Armed Forces once the Second World War broke out—a war in which many more Confessing Church members died fighting for Germany than in opposing the regime or the war—Niemöller lamented his earlier stand on the prewar repression of the Jews: "I did not realize that we would have to pay for these restrictions with our own liberty."[52]

Lest we think Bonhöffer and Niemöller were alone in their antisemitic attitudes, let us take the case of Karl Barth. Though Swiss and Reformed, Barth was the predominant Protestant theologian in Lutheran Germany during this period. In fact, he was perhaps the most influential Protestant theologian of the twentieth century. To his credit, his position on Christians of Jewish background was that there should be no limit whatsoever on their rights. Both the Church itself and the government, he argued, must treat these converts precisely as they treated any other Protestant, or else the Church could not itself be considered Christian. Yet Barth said little about humane treatment for Jews. Like Niemöller, his theology was Christological, not humanistic. It was precisely this emphasis that allowed him and many other Christians to avoid their moral responsibilities, especially when it came to the Jews. Barth's ethical *Weltresignation* left secular life and the fate of the Jews wide open to Nazi control. He wrote in 1943 that "the Church has no special duty to serve mankind nor the German people. It must only serve the word of God."[53] His attitude toward the Jews seemed to harden as the Holocaust grew worse. Although he told a Swiss audience on 5 December 1938 that in attacking the Jews the Nazis assault the very roots of the Christian Church, in 1942 he wrote that Jews deserved their fate.[54] Judaism was "outmoded and superseded." Jews had revolted against God and therefore experienced "the sheer, stark judgment of God. . . . The Jews of the ghetto . . . have nothing to attest to the world but the shadow of the cross of Christ that falls upon them."[55] For Barth, as for so many Christian theologians, the Jews were not real, individual people, they were bloodless symbols of those evil human beings who refused to love his Christian God. He candidly admitted that "the reason for the grudge we bear the Jew [is] that he is a mirror in which we see ourselves as we are, i.e., we see how bad we all are."[56] As late as 1953, Barth argued, "The existence of the synagogue beside the church is something like an ontological impossibility, a wound, yes a gaping hole, in the body of Christ Himself, that is simply unbearable."[57]

After the war, most of the members of the *Bekennende Kirche* felt guilt and contrition for their behavior during the Holocaust and publicly expressed their sorrow.[58] In October 1945, the Confessing Church proclaimed to the world at Stuttgart a "confession of guilt" that during Hitler's Reich they "had not confessed courageously enough, not believed cheerfully enough, not loved zealously enough."[59] In 1948, the National Brethren Council of the Protestant Church in Germany stated:

> What has happened, . . . we allowed to happen in silence. . . . Antisemitism rose and flourished not only among the people (who still seemed to be a Christian nation), not only among the intelligentsia, and in governmental and military circles, but also among Christian leaders. . . . The Churches had forgotten what Israel really is, and no longer loved the Jews. Christian circles washed their hands of all responsibility, justifying themselves by saying that there was a curse on the Jewish people. . . . We Christians helped bring about all the injustice and suffering inflicted upon the Jews in our country.[60]

But it was Martin Niemöller who was to make the most striking confession of guilt. In a lecture in Zurich on 7 March 1946, he stated, "Christianity in Germany bears a greater responsibility before God than the National Socialists, the SS, and the Gestapo. We ought to have recognized the Lord Jesus in the Brother who suffered and was persecuted despite him being . . . a Jew . . . . Are not we Christians much more to blame, am I not much more guilty than many who bathed their hands in blood?"[61]

The leading Roman Catholic prelate was Munich's Cardinal Michael von Faulhaber. During the anti-Jewish activities of the German government in April 1933, Faulhaber opined that there were more important issues for the Church than protecting Jews. He even declined to defend Jews converted to Catholicism since he felt that baptism gave no one leave to expect earthly advantage from it.[62] His position was enunciated more clearly in his Advent Sermons for 1933, where he pointed out that "Israel had repudiated and rejected the Lord's anointed, had . . . nailed Him to the Cross. Then the veil of the Temple was rent, and with it the covenant between the Lord and His people. The daughters of Sion received the bill of divorce, and from that time forth Assuerus [the Wandering Jew] wanders, forever . . . ."[63] In 1934, in response to international Christian comment that he had defended contemporary Jews, Faulhaber reaffirmed that anything positive about Jews in his 1933 Advent Sermons had referred only to Jews who lived before

Christ.[64] As late as November 1936, after nearly four years of Nazi rule, Faulhaber proclaimed that "the Reich Chancellor undoubtedly lives in belief in God. He recognizes Christianity as the builder of Western culture."[65] He told Hitler that "as supreme head of the German Reich, you are, for us, the authority willed by God, the legal superior to whom we owe reverence and obedience."[66] Faulhaber also defended "racial research and race culture," and argued that the Church had always recognized the importance of race, blood, and soil—though he warned against hatred of other races and "hostility to Christianity."[67]

Many other German Catholic prelates were strongly antisemitic. Bishop Hilfrich of Limburg, for example, argued that Christianity had not developed from the Jews, but instead had progressed in spite of these "God-killers." Cardinal Bertram agreed with Wurzburg Vicar-General Miltenberger that the Church had always recognized the importance of race, blood, and soil. The same perspective had been expressed by Monsignor Groeber, archbishop of Freiburg, who himself had joined the Nazi SS in 1933 as a "promotive member" and had publicly argued for "the national right to maintain unpolluted its racial origin and to do whatever necessary to guarantee this end."[68]

A minority of German Catholics opposed the regime. Jesuit Fr. Rupert Mayer, whose position was that one could not be a Catholic and a Nazi at the same time, was disappointed in Cardinal Faulhaber's not speaking out for the Jews.[69] A Jesuit critic of Bishop Berning's visit to Hitler, Friedrich Muckermann commented: "We face the shocking truth that the only word that a German bishop until today has publicly said about the barbarities of the concentration camps is a word of glorification of Hitler and of a system that has brought about these barbarities."[70] The Church later squelched Muckermann's criticisms.[71] In the wake of the Nazi Kristallnacht pogrom in 1938, Dean Bernhard Lichtenberg, provost of Hedwig's Cathedral in Berlin, deplored Nazism and specifically condemned the German attack on the Jews, for whom he publicly prayed. "What happens today we have witnessed; outside the synagogue is burning, and that also is a house of God."[72] In 1943, Alfred Delp, a German Jesuit who would later pay with his life for his involvement in an assassination attempt on Hitler, asked at the annual German Catholic Bishops' Conference: "Has the Church forgotten that it must every now and then say *you must not*? Has the Church lost sight of the Commandments? . . . Has the Church forgotten human beings and their fundamental rights?" The Jesuit order split in its support or opposition to Hitler.[73] In 1937, Berlin Bishop Konrad von Preysing wrote a pastoral letter attacking the Hitler regime, accusing the government of violating German consciences.[74] In 1943, von Preysing threatened to resign over the collaborative behavior of the other German bishops, and he unsuccessfully urged Pope Pius XII "to issue an appeal in favor of the unfortunate [Jews]."[75]

Christian Resistance to the Third Reich

There was a Protestant opposition in Germany to the Hitler regime, but it was halting, usually disorganized, and almost always ineffective. There were several reasons for this. First, Protestants, as well as Catholics, feared with good reason that open opposition could destroy the Churches. Second, there was hesitation among these religious to perform the "illegal" and "immoral" acts—from dissembling to murder—that real resistance required. Third was the influence of Luther's notion of the Two Kingdoms, which many contemporary Lutherans interpreted to mean that they had no right to criticize the Nazi government, for it had exclusive authority in the secular, political arena of German life. Fourth, German Protestants were, on the whole, nationalistic and antisemitic, and much of what the National-Socialists said and did in both foreign policy and domestic politics seemed logical and congruent with their own values, especially the regime's attacks on modernism, democracy, Communism, and the *éminence grise* they saw behind all these movements, "the eternal Jew." The Nazi attack on the Jews was at the heart of the Hitlerian *Weltanschauung*, and helping the Jews would have therefore been tantamount to the most radical kind of dissent and resistance to the regime and its ideology.[76] This was the very step that most German Protestant leaders and laymen were unwilling to take.

Most of these resisters excluded Jews from their domain of the decent. Jews were indecent, and that is how most of the best and brightest Germans treated them. Like Heidegger and most Germans, the resisters were willing to accept the Third Reich's Final Solution to the Jewish Problem.

Many contemporary historians have remarked on this situation. Klaus Scholder, historian of the Church Struggle in the Third Reich, has written that in 1933 most churchgoing Protestants accepted Hitler's national revolution, and along with it, the persecution of the Jews. And he noted the overall "confusion, blindness, despondency, and weakness of the Protestant Church [in Germany]."[77] Others have noted that the Churchgoing *Mittelstand* objected to Nazi methods but saw little wrong with antisemitism itself and that the anti-Hitler resistance in Germany, and the German Churches to which it was closely tied, like most Germans, was never really concerned with the fate of the Jews. A broad consensus supported Nazi values.[78]

Their perceptions already distorted by religious antipathy, many Germans may have falsely imagined, as the U.S. president Franklin Roosevelt stated at the Casablanca Conference in 1943, that "the number of Jews engaged in the practice of the professions . . . should be definitely limited[; this] would further eliminate the specific and understandable complaints which the Germans bore toward the Jews in Germany, namely, that while they represented a single part of the population, over 50 per cent of the lawyers, doctors, school teachers, college professors, etc., in Germany were Jews."[79]

Without this wide German antisemitic consensus, as Robert Gellately has concluded, the Nazi regime's "antisemitic policies would have remained so many idle fantasies."[80] The major disagreements most Christian Germans had with the Nazis with regard to the Jews were the Churchgoers' squeamishness in facing public anti-Jewish violence within Germany. The Reich government solved this "problem" by siting the death camps away in Poland—for many Germans, the *anus mundi*, "the place at which all the world's excrement is unloaded,"[81] or, to put it into today's vernacular, the "asshole of the world." The German concentration camps were simply ignored, behind their walls, as something "nicht schoen," not very nice.[82] The reticence of most of the young anti-Nazi idealists of Helmut von Moltke's Kreisau Circle, for example, in concerning themselves with the Jews may be mostly traced to their "strong bonds with Christianity, both Lutheran and Catholic, which they regarded 'as the foundation of the moral and religious revival of our people.' "[83]

Most conservative German resisters were antisemitic—Claus von Stauffenberg, Carl Goerdeler, Adam von Trott, and the Kreisau Circle. They believed that Germany was and should be based on Christian values. Like the Nazis and other Germans, most resisters regarded the Jews as unwanted evil aliens, never to be permitted German citizenship. These resisters wanted a "volkish Fuehrer state without Hitler."[84] As Bonhöffer's friend Eberhard Bethge put it, "we were against Hitler's church policy, but at the same time we were antisemites."[85] John Weiss points out that only a small minority of the resisters, such as the White Rose and Helmut von Moltke, were prepared to accept Jews in their Germany.[86] Of those Germans who were arrested for their opposition to the government, only 20 percent even mentioned the Jews. Exceptions worthy of note were members of the White Rose—symbol of the moral good of Christianity. This resistance group was led by Munich university students Hans and Sophie Scholl and philosophy professor Kurt Huber, righteous Christians, who publicly protested against the inhumanity of the Hitler regime.[87]

Christianity was respected in Germany both within and outside the Nazi party. The Nazis claimed their goals were the same as Christianity's; they were anticlerical in that they opposed the churches as potential powerbases outside Nazi authority. Most Protestants, whether Nazis themselves or anti-Nazis, still held to traditional Christian antisemitism—as did most members of the Catholic Church in Germany. A vision of how a victorious Third Reich would have treated Christian institutions is revealed in regulations concerning the Warthegau, the territories annexed into the Reich that were conquered from Poland and under complete control of Bormann's Party Chancellery. Here the churches were no longer state churches but voluntary institutions with no connection to German churches or to the Vatican. There were no church taxes collected by the Reich and no youth or welfare groups. Church property was limited to the church buildings themselves. Poles and Germans were to use separate churches and ministers had to have other occupations in addition to their ministry. The point is that these Christian institutions still existed and did so independently of the Reich government. In contrast to the razing of church buildings in the Soviet Union, in Nazi Utopia, the churches were allowed to exist.[88]

Because antisemitism was the norm, most Germans accepted the anti-Jewish policies of the Third Reich before the war and the mass murders of the wartime Final Solution as a *fait accompli*. That is, both the Nazis and the vast majority of Christian Germans, by their collaboration or silence, seemed to agree that the Jews should suffer discrimination, expropriation, or expulsion from society as part of "God's plan" for them—and finally, disappearance into the night and fog of the East once the war began. The Reich's antisemitic legislation was very popular and widely supported.[89] In contrast to their acceptance of anti-Jewish measures, many Germans protested the Nazis' brutal treatment of Polish Catholic slave laborers in Germany.[90] In April 1942, Alfred Rosenberg's Foreign Policy Bureau reported to the Ministry of Eastern Occupied Territories: "Evidently the Polish question can not be resolved simply by the liquidation of the Poles, as is the case with Jews. Such a solution of the Polish question would burden the Germans people far into the future, would obliterate all sympathy for it, at the same time causing other people to wonder whether they might not have to undergo the same fate in due time."[91] As Ayçoberry convincingly demonstrates, despite individual acts of courage, scattered, disorganized opposition could never compete with the enthusiastic support or the passivity of the majority of Germans.[92]

Hitler and Christian Antisemitism

Adolf Hitler served as the catalyst, as the "drummer," in Joachim Fest's term, inciting the already antisemitic German masses, attracting not only

Germans but also most of Europe and even elements among the Allies to his war against the Jews. Hitler served as a leader who led a movement that integrated traditional anti-Jewish Christian ideas (including Christian racism) and policies with a nationalist-racist ideology that did not flinch from instituting a program of genocide of an unprecedented scale and "effectiveness." As "Führer" of the German Volk and a "man of destiny," he also provided a political sanction to the religious antisemitism deeply ingrained by the Church into the average Christian, who regarded the Jew with disdain, or worse. Hitler helped transform Christian anti-Jewishness into mass murder of Jews on an unprecedented scale and more efficiently than ever before.

Although many Germans feared Hitler, many others admired his use of force against Germany's "enemies" outside and inside the nation.[93] Although he planned ultimately to destroy Christian institutions,[94] Germans believed Hitler when he promised to employ the power of the Third Reich to end disunity and create a Germanic and Christian nation.[95] Germany did not have a commanding Christian voice insisting, "First a human being, then a Christian: this alone is life's order"—the motto of Denmark's leading Lutheran humanist Nikolai Grundtvig.[96] Instead, the paradigmatic German-Christian was Martin Luther. There were few Christians who sensed that Jews were part of the German polity, few who felt that Jews were authentic human beings.

Christian antisemitism influenced Hitler's ideas and behavior. When Hitler was a child, when most learn their prejudice, he learned antisemitism from his Austrian environment. Indeed, Austria was the Western nation most influenced by Christian antisemitism.[97] As in most European nations, Austrian Catholics regarded the Jews as cursed aliens, deniers of Christ, "the veritable offspring of the devil."[98] Indeed, the *Anschluss* of 1938 released such powerful anti-Jewish emotions that the Austrians had to be restrained by the *Germans* from attacking Jews.[99]

Hitler and other Nazi leaders—some having previously served as ministers, church officials, and theology students—never failed publicly to give the impression that they were pious and devout Christians.[100] Although most Nazis came from a Protestant background, a disproportionate number of the leaders of the Final Solution were Catholics. Austria, the seedbed of Catholic antisemitism, gave birth to Adolf Hitler, Adolf Eichmann, Ernst Kaltenbruner, Odilo Globocnik, Rudolf Hoess, and Franz Stangl. Other non-Austrian Catholics were Josef Goebbels, Heinrich Himmler (the young Himmler had written, "Come what may, I shall always love God, pray to Him, and adhere to the Catholic Church and defend it, even if I should be expelled from it"; he did not formally leave the Church until 1936),[101] Reinhold Heydrich, Julius Streicher. After the First World War, Austrian and German Catholics, often led and supported by their priests, established numerous antisemitic paramilitary groups. In the last election of the Hapsburg Empire, more than two-thirds of German Austrians voted for candidates who were extreme antisemites. Austrians joined the Nazi Party and the SS at a rate almost double that of Germans. Less than ten percent of the German population, Austrians comprised nearly half the concentration-camp staffs.[102] Austrians may have been responsible for half of all war crimes.[103] The Austrian bishop Gfollner of Linz in his pastoral letter of January 1933 stated that although one could not be a good Catholic at the same time as being a Nazi, it was the duty of all Catholics to adopt a "moral form of antisemitism. . . . Our modern society . . . should . . . provide a strong barrier against all the intellectual rubbish and moral slime which, coming largely from Jewry, threatens to flood the world."[104]

In Austria it was more than likely that from his youth onward Hitler was taught about Christian antisemitism. The young Hitler was impressed by the masses and sermons at the local Catholic church. Besides, for two years he received religious instruction at the choir school of the Benedictine monastery at Lambach, where he served as altarboy and choirboy. At this time he developed the ambition of perhaps becoming a priest himself—an intention approved by his devout mother and even his anticlerical father. As he wrote in *Mein Kampf*, "I had an excellent opportunity to intoxicate myself with the solemn splendor of the brilliant church festivals. As was only natural, the abbot seemed to me, as the village priest had once seemed to my father, the highest and most desirable ideal."[105]

Despite Hitler's own assertions that his antisemitism occurred later, Hitler must have learned it as a youth. If the environment of Hitler's youth was typical of the time and place, it must have been rife with traditional religious antisemitism, and if the attitudes toward Jews of the abbot Hitler so admired were common for the time, then they were anti-Jewish. In Linz, when Adolf was a boy, Germans publicly demonstrated against Jews. Alois, Hitler's father, and mother were most likely antisemitic. The only teacher he admired was an antisemitic historian and politician.[106] As a schoolboy, Hitler passed out antisemitic literature. At the Staatsrealschule in Steyr, Hitler did have further contact with Christian teachings. He took religion classes, where his grades ranged from adequate to satisfactory.[107] As a teenager, he commented to his friend Kubizek as they passed Linz's small synagogue in the Bethlehemstrasse, "That does not belong here in Linz." Moreover, although Hitler talks about race *ad nauseam*, he nevertheless wrote in his autobiography that of the few Jews who lived in his hometown it was their religion, not their race, that distinguished them from the other German Austrians. "I saw no distinguishing feature but the strange religion."[108]

Despite his denials, Hitler most likely arrived in Vienna already a Christian antisemite. In Vienna he was struck by the numerous eastern European Jews in Orthodox dress and manners.[109] Hitler wrote: "Among our people the personification of the devil as the symbol of all evil assumes the living shape of the Jew."[110] This image was Christian, not just Hitlerian or Nazi. This kind of antisemitism spread wherever Christianity spread.

In Vienna, Hitler was also able to familiarize himself with the Christian antisemitism of Karl Lueger—the leading Austrian antisemitic politician at the end of the nineteenth and turn of the twentieth century. Lueger was mayor of Vienna, representative to the Austrian Parliament, and head of the Christian-Social Party. Although Hitler objected to the Christian-Social's "half-hearted" antisemitism, he wrote that "If Dr. Karl Lueger had lived in Germany, he would have been ranked among the great minds of our people."[111] It was on the twenty-eighth anniversary of Lueger's death that Hitler ordered the German invasion of Austria.

Hitler was also aware of the history of theological antagonism to the Jews and privately he expressed his admiration for the anti-Jewish ideas of "all genuine Christians of outstanding calibre." He mentioned John Chrysostom, Pope Gregory VII (who, along with Revelations and Pope Pius IX used the phrase "synagogue of Satan"), Thomas Aquinas,[112] Goethe, Father Rohling, Heinrich Treitschke, Richard Wagner,[113] and, especially, Martin Luther as such Christians.[114]

Like Paul de Lagarde, Hitler regarded Luther as "one of the greatest Germans," "the mighty opponent of the Jews," "a great man, a giant," who had found himself, as Hitler had, in his antisemitism.

> He saw the Jew as we are only now beginning to see him today. But unfortunately too late, and not where he did the most harm—within Christianity itself. Ah, if he had seen the Jew at work there, seen him in his youth! Then he would not have attacked Catholicism, but the Jew behind it. Instead of totally rejecting the Church, he would have thrown his whole passionate weight against the *real* culprits.[115]

In 1932, Hitler, speaking informally in his Munich flat, observed that "Luther, if he could be with us, would give us [National Socialists] his blessing."[116] Hitler's "Final Solution to the Jewish Problem" paralleled Luther's program for the Jews in almost every respect—the destruction of Jewish culture, economy, and sociopolitical standing, expulsion, and mass murder.

One of Hitler's most quoted lines from *Mein Kampf* was: "In defending myself against the Jews I am acting for the Lord."[117] This phrase appeared on calendars and posters displayed all over Germany. Between 17 and 23 May 1936, it served the SS as their Motto of the Week.

Hitler was aware of the Christian ritual-murder defamation. Between 1880 and 1945, there were as many instances of the ritual-murder defamation against Jews as during the entire Middle Ages.[118] Many areas of Germany and Austria revered local saints as "martyrs of the Jews." In the 1920s and 1930s, Julius Streicher's Nazi journal *Der Stürmer*—the most popular of Nazi publications and Hitler's personal favorite—was filled with references to ritual murder and other religious accusations against Jews, in particular their association with the devil. "Historical documents [and] works of art portraying these fables in individual Catholic churches prove that what was written about really happened, indeed that the Catholic Church recognizes the reality of ritual murder."[119] In 1926, *Der Stürmer* published a story and cartoon on Jewish ritual murder. The cartoon showed three Jewish men drinking blood from a slaughtered blonde Polish woman. *Der Stürmer*'s famous ritual-murder issue of 1 May 1934 contained many articles on the subject and a front-page drawing of stereotypical Jews catching the blood from the severed veins of blonde women and children who were hanged upside down.[120] After an international uproar, Hitler banned this issue of *Der Stürmer* on the grounds that Streicher's comparison of the Christian sacrament of communion with Jewish ritual murder was an insult to Christianity.[121] On 25 December 1941, *Der Stürmer*'s Christmas message read: "To put an end to the proliferation of the curse of God in this Jewish blood, there is only one way: the extermination of this people, whose father is the devil."[122]

In a private conversation with Dietrich Eckart, Hitler's closest friend up to the 1923 Munich *Putsch*, Hitler observed that "the Jews had continued to perform ritual murders" until recently.[123] In 1935, Hitler's government permitted the republication of Luther's *The Jews and Their Lies*, which contained references to Jewish ritual murder and the Christian's obligation, in turn, to murder the Jews.[124] Under the Third Reich, Luther's birthday was celebrated, there were Martin Luther schools, Martin Luther monuments, Martin Luther streets, Martin Luther churches.[125] Presumably with Hitler's knowledge,[126] in 1943, Himmler ordered Ernst Kaltenbrunner, his chief subordinate in the SS, head of the Reich Central Security Office to discover cases of Jewish ritual murder "wherever Jews have not yet been evacuated," notably in England, Romania, Hungary, and Bulgaria, and publicize them.[127]

Hitler's public speeches and private conversations indicate that Christian antisemitism inspired many of his anti-Jewish ideas, which paralleled those of his listeners, much of it from the Gospels,[128] and he referred to several Roman Catholic antisemites along with Martin Luther. Throughout the 1930s, Hitler's public statements and private conversations contained biblical allusions.[129] He sometimes expressed himself, as in this speech on 12 April 1922, in the words of the Gospels themselves.

> I would be no Christian . . . if I did not, as did our Lord two thousand years ago, turn against those by whom today this poor people [Christian Germany] is plundered and exploited. . . . My feeling as a Christian points me to my Lord and Savior as a fighter. It points me to the man who . . . recognized these Jews for what they were and summoned men to fight against them . . . . In boundless love, as a Christian and a human being, I read through the passage which tells us how the Lord rose at last in His might and seized the scourge to drive out of the Temple the brood of vipers and adders. How terrific was His fight against the Jewish poison. Today, after two thousand years, with deepest emotion I realize more profoundly than ever before the fact that it was for this that He had to shed His blood upon the Cross.[130]

In Hitler's chapter "Nation and Race," in his book *Mein Kampf*, Hitler called Judaism a "monstrous" religion and the Jew a "product of . . . religious education." Echoing the ideas of the Church Fathers, Hitler observed that the Jew's

> life is only of this world, and his spirit is inwardly as alien to true Christianity as his nature two thousand years previous was to the great founder of the new doctrine. Of course, the latter made no secret of his attitude toward the Jewish people, and when necessary he even took to the whip to drive from the temple of the Lord this adversary of all humanity, who then as always saw in religion nothing but an instrument for his business existence [this-worldly attitude]. In return, Christ was nailed to the cross . . . .[131]

On 1 February, 1933, in his first address to the nation after he took power, Hitler stated that "the national government . . . will take Christianity as the basis of our collective morality . . . ."[132] Later in the month, repeated in Koblenz the next year, he told a Stuttgart audience that "Christians . . . now stand at the head of Germany. I do not merely *talk* of Christianity, no, I also profess that I will never ally myself with the parties which destroy Christianity. . . . We wish to fill our culture once more with the spirit of Christianity—and not only in theory. No, we want to burn out the symptoms of decomposition in . . . our whole culture . . . ."[133]

To the Reichstag in March 1933 he stated that "the two Christian Confessions [are] the weightiest factors for the maintenance of our nationality. . . . [T]he struggle against materialistic [read, Jewish] views and for a real national community is just as much in the interest of the German nation as in that of the welfare of our Christian faith."[134] Germans believed

Hitler when he promised to employ the power of the Third Reich to end disunity and create a Germanic and Christian nation.[135]

Hitler spoke on 26 April 1933 with two Catholic fellow travelers, Bishop Wilhelm Berning of Osnabrueck and Mgr. Steinmann. Bishop Berning, a member of Goering's State Council, in June 1936 visited a number of concentration camps to remind the prisoners "of the duty of obedience and fidelity towards people and state that was demanded by their religious faith." He also praised the guards for "their work in the camp" and ended his visit with three *Siegheils*. He was raised to archbishop in 1949.[136] Steinmann was the Berlin vicar-general who greeted Catholic faithful with "Heil Hitler."[137] To these prelates, Hitler acknowledged that he was doing to the Jewish "parasites" what the Church had been doing to them for 15 centuries. In the context of the early Third Reich's attacks on Jews—riots in March, Aryan paragraph in April that disregarded the Christian status of Jewish converts to Christianity—Hitler began: "You have attacked me because of [my] handling of the Jewish question. [But] the Catholic Church has regarded the Jews as parasites and banished them to ghettos and so forth, so you know what the Jews are like. . . . I discern parasites in the representatives of this race for state and church, and perhaps I am doing Christianity the greatest service." We have no report as to what the prelates replied, if anything. But they later described the talks as "cordial and to the point."[138]

Two days after the Berning–Steinmann–Hitler meeting, papal nuncio Orsenigo reported to Vatican Secretary of State Pacelli that "the social elimination of the Semitic element continues on a large scale." In May 1933, Orsenigo reported a conversation he had with Hitler in which the German leader told him that "neither a private life nor a state . . . could be imagined without Christianity." Orsenigo expressed no doubts about Hitler's sincerity, nor did he add any comments himself on actions being taken by the government against the Jews. Orsenigo seemed alarmed by only two issues: Bolshevism and Protestantism.[139]

Hitler's speech to political leaders of the Nazi party at Nuremberg in 1936 included "an astonishing montage of Biblical texts," especially from the Gospels of Matthew and John.[140] In 1938, in a conversation with Minister of Justice Hans Frank, Hitler noted, "In the Gospel, when Pilate refuses to crucify Jesus, the Jews call out to him: 'His blood be upon us and upon our children's children.' Perhaps I shall have to put this curse into effect."[141] On 24 February 1939, three weeks after his Reichstag speech threatening the Jews with destruction should war start, Hitler stated, "Today the Jewish question is no longer a German problem, but a European one." The German people's needs must be satisfied by Jewish expropriation. In this sense, "we are true Christians!"[142]

Hitler summed up his attitude in his famous 30 January 1939 Reichstag speech: "The German priest as servant of God, we shall protect; the priest as political enemy of the German state, we will destroy." Though some Nazis planned to destroy Christianity itself, many others, like Hitler himself, admired Jesus, especially because he saw him as having fought against their common enemy, the Jews; Hitler believed in a personal God; Hitler felt that the churches would always be a crucial part of Germany's political life; Hitler praised the Ten Commandments; he dismissed paganism; he realized he needed the churches to maintain the support of the German people.[143]

Hitler considered himself, at least for public consumption, as a Roman Catholic. The Austrian Church taught its parishioners, as Austrian Bishop Keppler wrote, that the Jews "are a thorn in the flesh of Christian peoples, suck their blood, enslave . . . , and contaminate [their] culture and morality."[144] Hitler participated in regular Communion for the first 30 years of his life.[145] Until the very end in 1945, he continued to allow the state to withdraw a tithe for the Catholic church from his salary without ever publicly indicating that he was a man without belief in the Catholic faith.[146] Even after he was in power, he often times made gifts to small Church congregations. His public image was always that of "a religious man interested in the church."[147] Hitler wrote that "the religious doctrines and institutions of the leader's people must always remain inviolable."[148] From the Church's point of view, Hitler's works have never been put on the Index of Forbidden Books, nor has he, or any other Catholic who participated in the "Final Solution," ever been officially excommunicated from the Church for war crimes.[149]

Christian beliefs were not seen as inconsistent with Nazi ideology nor with state antisemitism.[150] The Protestant and ultimately the Catholic leadership endorsed the regime wholeheartedly. Just as apples do not fall far from the tree, so the Protestant and Catholic parishioners followed their pastors' antisemitic lead.[151] The majority of both Protestants and Catholics endorsed the Nazi regime (even though in the March 1933 elections the Nazis received only 44 percent of the vote, many who voted for other parties still sympathized with the Nazis' policies and attitudes toward the Jews),[152] and yet only a minority of Germans personally adopted a secular racist antisemitism.[153] Two weeks after the Reich law of 14 July 1933 had ordered ecclesiastical elections, fully two-thirds of German Protestants joined the pro-Nazi German-Christian movement (Faith Movement of German-Christians, or *Glaubensbewegung Deutschen Christen*) or voted for it. Hans Schemm was a Protestant schoolteacher, Nazi party member of the Reichstag, and a founder of the German-Christian movement. He proclaimed that "Our politics are Germany and our religion Christ."[154] At the

1958 *Einsatzkommando* trials at Ulm, the Protestant pastor of the unit explained why he had been a silent witness to atrocities against Jews: "These acts were the fulfillment of the self-condemnation which the Jews had brought upon themselves before the tribunal of Pontius Pilate."[155] Christopher Browning has examined German killers who were members of the Reserve Police Battalion 101. Although he does not specify Christian antisemitism and racism, he points out that the "constant, pervasive, and relentless" antisemitism and racism that conditioned all Germans, dovetailing with the dehumanizing effects of war, led to atrocity.[156]

The Nazis, the Third Reich, and the SS could not have achieved its genocidal goals without the acquiescence of the Churches and the voluntary cooperation of millions of Germans and other Europeans.[157] The German Churches had to open their baptismal records; the government's bureaucracy, the railroads, the physicians, the Foreign Office, the K-12 teachers, the university professors, the scientists, and so forth—all Christians—had to cooperate to complete the final solution. The Gestapo itself would have been ineffective if the general public had not voluntarily collaborated with government officials.[158] Nazi pamphlets urged all Christians to join the party in order to conclude the bimillennial Christian Crusade against the Jews.[159] Hitler purged Nazi Party members who denounced the Churches.[160] When Christians protested against government policies, even in the middle of the war, even in front of Gestapo headquarters, even to save Jewish spouses, the Nazis backed down. Goebbels decision to return about 10,000 Jews who had been arrested back to their Christian Aryan families, not to send them off to Auschwitz, was confirmed a week or so later by Hitler himself.[161] What if this kind of German protest had been as widespread as German collaboration with the Nazis? Would not the Final Solution been stopped dead in its tracks?

Like Josef Goebbels, Streicher was nominally Catholic and, more important, saw in Christianity "one of the greatest anti-Jewish movements." Of Jesus and his disciples, only Judas was Jewish. "The crucifixion of Christ is the greatest ritual murder of all time."[162] Streicher referred to the early Holocaust as "Jewish punishment for Golgotha."[163] Church Father Cyprian's assertion that "The Bible itself says that the Jews are an accursed people . . . the devil is the father of the Jews" became in 1936 the masthead of Streicher's *Der Stürmer*—Hitler's favorite reading.[164] In 1935, his speech to the Hitler Youth could have been delivered by most of the Church Fathers *mutatis mutandis*, without compunction. In many ways, Streicher's speech was less vitriolic than the anti-Jewish sermons of John Chrysostom or the writings of Martin Luther—antisemitic Christian writers to whom

Hitler himself refers.[165]

> Boys and girls, . . . only one people remained victorious in this dreadful war [First World War], a people of whom Christ said its father is the devil. That people had ruined the German nation in body and soul. [Once Hitler had come to power] the human race might be free again from this people which has wandered about the world for centuries and millenia, marked with the sign of Cain. . . . A chosen people does not go into the world to make others work for them, to suck blood. . . . It does not go among the peoples to make your fathers poor and drive them to despair. A chosen people does not slay and torture animals to death. . . . Boys and girls, for you we went to prison. For you we have always suffered. For you we had to accept mockery and insult, and became fighters against the Jewish people, against that organized body of world criminals, against whom already Christ had fought, the greatest antisemite of all times.[166]

Randall Bytwerk concluded that Streicher was speaking "in a tradition of ancient familiarity." His antisemitism may have been more intense than most Germans', but his ideas on the Jews "were fundamentally respectable to them."[167] Average Germans, who bought Streicher's Nazi newspaper *Der Stürmer* felt very comfortable with the basic anti-Jewish values it embodied.[168] Just as the paper was posted in German towns and villages for all to read for free, so it was displayed in concentration-camp guard living quarters.[169]

During the war, German soldiers murdered Jews—even Jewish children—according to their own testimony because they saw the satanic Jews as a threat. At his postwar trial, similar to Adolf Eichmann and Rudolf Hoess, Otto Ohlendorf, commander of Einsatzgruppe D, explained that "all Jews including the children were considered to constitute a danger."[170] Following traditional Christian defamations against Jews, the Third Reich told its soldiers that Jews were the original historical murderers of the righteous, ritual murderers of innocent Christians—the *Ewige Jude*, the Eternal Jew, Germany's Christian and racial enemy.[171]

German-Christians emphasized Christian supremacy over the Jews, not the universal humane love of neighbor espoused by Jesus of Nazareth. In contrast to authentic *caritas*, these religious affected a superficial brotherly love restricted to fellow members of the Church.[172] When Christians heard, thought, or felt, "Love thy neighbor," they defined neighbor to exclude the alien Jew. At best, Jews were regarded as faded symbols, stereotypes, hardly human at all; at worst, as satanic deicides.[173] Most Germans did not perceive the Jews as *unserer*, ours. Even before the Holocaust, the Jews were suffering, according to Richard Rubenstein, a kind of "death in life."[174] Thus the Nazis, as well as most of their Gentile opponents and victims, saw the Jews not as real people being discriminated against, expropriated, and sent off to prison and concentration camps. Instead, they saw the Jews as having been condemned "by some Supreme World Court."[175] The Jews had, indeed, been condemned by the theologians and leaders of the Churches, the highest courts in history. This theologically generated sentence for the Jews long before Hitler's rise had "left the Jews helpless when the murderers came to collect them for the furnaces."[176] The German state became a murderous society once Hitler added political sanction to the preexisting religious hostility toward the Jews.

The great moral tragedy for both the Protestant and Catholic Churches was that most of the faithful and most of their bravest and brightest heroes were men and women who suspended their own standards of decency when it came to the Jews. In many ways they admired Hitler and felt comfortable with his policies: his anti-Bolshevism, his prosecution of the war, and his persecution of the Jews, so long as it was not in their backyards.

Racism or Religious Antisemitism

Following Hitler's lead, the Nazis were intent on demonstrating to the world and to the German people that they ran a "Christian" government. Even though the Nazi regime caused the deaths of hundreds, perhaps thousands, of Christian pastors, priests, and nuns, and millions of Slavic Christians, it seldom desecrated their churches and the sacraments.[177] In contrast, years before Kristallnacht, between 1923 and 1932, 128 Jewish cemeteries and 50 synagogues had been desecrated.[178] Himmler did not want clergymen in the SS, but they were asked to leave "in the most tactful and honorable fashion possible."[179] Yet no matter how hard the SS leadership tried, deChristianization never worked in the SS. One-quarter of SS members were practicing Catholics, to whom the Church never denied communion despite their participation in the Final Solution.[180] Even during wartime, Protestant and Catholic members of the Waffen-SS could receive their Churches' sacraments.[181] A small but symbolic example of the Reich's attitude toward Christian culture occurred at the site of the most active concentration camp in Austria, Mauthausen. While murdering tens of thousands of Jews there, the Nazis carefully preserved a medieval fresco portraying "Christ as judge of the world, the lamb of God." The Reich provided "for the drying of the already strongly endangered painting, and did so in the middle of wartime."[182]

Anti-Jewish biological racism was often expressed in public. Many rank-in-file Nazis regarded Alfred Rosenberg's bible of biological racism, *The Myth of the 20th Century*, with an "official awe" second only to that of Hitler's *Mein Kampf.* Its main themes were: Not social class, not economic conflict, but blood and race were decisive factors determining art, science, culture, and the course of world history; the German nation represented the "master race" of Aryans, whose mission was to subdue the supremely dangerous Jews, who were a parasitic, biological "germ" that threatened the purity of the Germanic aryan race. But within the antisemitic Nazi inner circle, including Hitler, these racist ideas were often ridiculed.

Although Hitler attempted to close the traditional religious loophole of baptism as an act that could save a Jew's life—Hitler worried that "if the worst came to the worst, a splash of baptismal water could always save . . . the Jew"[183]—a strictly biological racism was not as dominant in Nazi thought as earlier interpretations of the Third Reich assumed. Rosenberg himself, for instance, spoke of race as a "mythischen Erlebnis," "mythic experience."[184] His particular brand of *Judenhass* sometimes understood Judaism as a flawed religion and the Jews as an inferior religious community—he critiqued Jewish theology as causing the Jews' antisocial character—other times as a biological race.[185] Rosenberg held that the Jewish spirit was essentially immoral and therefore required a multiplicity of "moral laws" to control it.[186] Besides, an appreciation of Rosenberg's biological-racist ideas appears to have represented only a minority attitude among the Nazis and certainly among the German people. The *Mythus* was little understood and "read primarily by National Socialist sub-ideologists in search of slogans of their own and by opposition ideologists seeking weak points in the Nazi armor."[187] Von Papen reported in his *Memoirs* that Hitler had ridiculed Rosenberg's book in no uncertain terms.[188] Hitler told Bishop Berning on 11 April 1942, "I must insist that Rosenberg's *The Myth of the 20th Century* is not to be regarded as an expression of the official doctrine of the party. . . . It gives me considerable pleasure to realize that the book has been closely studied only by our opponents. Like most of the Gauleiters, I have myself merely glanced cursorily at it." Goering and Goebbels thought Rosenberg was ridiculous.[189] Goering reportedly called Rosenberg's *Mythus* "junk," and Goebbels described it as "*weltanschauliches Gerülpse*," "philosophical belching."[190] Goebbels also spoke of the "rubbish of race-materialism" and regarded biological racism as without resonance for National-Socialism's understanding of itself. He dismissed Himmler's ideology as, "in vielen Dingen [für] verrückt," "in many regards, mad."[191] G.M. Gilbert, the psychologist who examined the Nazi Nuremberg defendants—the highest ranking Nazi leaders, aside from those who killed themselves or escaped—reports that none of them had ever read Rosenberg.

Heinrich Himmler adhered to the worship of medieval Germanic ancestors and *Blut und Bogen*, blood and soil. (Hitler and Goebbels, among other leading Nazis, ridiculed Himmler's beliefs.)[192] But Himmler himself defended Christianity. "In ideological training," he commanded, "I forbid every attack against Christ as a person, since such attacks or insults that

Christ was a Jew are unworthy of us and certainly untrue historically."[193] (Hitler certainly agreed, when on 21 October 1941, he spoke of "the Galilean, who later was called the Christ . . . .He must be regarded as a popular leader who took up His position against Jewry. . . . It's certain that Jesus was not a Jew.")[194] Privately and publicly, Himmler respected Christianity itself.[195] Even within the SS, Himmler forbade "SS members to pester, annoy or mock another due to his religious views . . . the religious convictions of his neighbors [are] holy and inviolable."[196] Going beyond this, Himmer "most strictly [forbade] any disturbance as well as any tactlessness regarding religious events of all confessions (i.e., processions of the Catholic Church). Likewise, a tactful deportment when churches are visited . . . goes without saying."[197] Not only was grace said at his house before meals by his daughter and in Himmler's presence, but he also told his friend Felix Kersten in the 1940s that "I've nothing against Christianity in itself; no doubt it has lofty moral ideas." In May 1944, to SS leaders, Himmler again expressed his approval of the Catholic Church but his hostility to its seeking secular power: "[If] the Christian Catholic Church had remained what it was, fulfillment to the soul, the mediary to the Lord . . . that would have been fine." But the Jews had corrupted the Church in order to ruin the German *Volk*. This is what was wrong with the Church.[198]

Another leading SS anti-Christian was Reinhard Heydrich. But he reserved his vituperation for the institutions of Catholicism as fighting Hitler's plan to bring religious harmony to Germany.[199]

Secular racist antisemitism served as an ideological justification necessary only to those Nazis who saw themselves as anti-Christian. A purely secular, unChristian racism would have appealed in the end only to the few anti-Christian Nazis who needed a nonreligious justification for their antisemitism.[200] As Christof Dipper has recently observed, "dilettante racial theories helped give modern antisemitism the scientific veneer [traditional Christian antisemitism] lacked . . . ."[201]

The Church had made the Jews stateless beings long before the Nazis had in their Nuremberg Laws; just as in Germany, there had been no legitimate place for the Jews in the medieval *societas Christiana*, the mystical corpus Christi of the Church. The Church had also established administrative and legal precedents for handling Jews that served the Nazis as a model. The governments of France, Poland, Hungary, and Romania also issued religiously based anti-Jewish legislation during the Holocaust.[202]

Crucial to the Nazis' legal definition of Jew was the matter of religion.[203] The anti-Jewish discriminatory decree of April 1933 and the Nuremberg Laws of 1935 used the religion of the grandparents and Jewish descent therefrom as the basic criterion of Jewishness. The Nazis had always before them the example of dozens of Church laws hostile to the Jews and hundreds more religiously inspired secular regulations that discriminated against Jews. Hitler's lawyer Hans Globke found specific historical precedent for making Jews wear a yellow star in the records of the Catholic Church's Fourth Lateran Council of 1215.[204] In 1941, Vichy Ambassador Léon Bérard also noted medieval Christian precedents, along with a summary of Thomas Aquinas' theological position on the Jews, in his defense of Vichy's discriminatory laws.[205] Vichy High Commissioner for Jewish Affairs Xavier Vallat argued that Vichy's antisemitic legislation was based on "precedents in the historical past of our own nation and in that of Christianity." In their postwar trials, both Bérard and Vallat cited Thomas Aquinas' work as justifying Vichy's Jewish policy. Vallat referred to "the anti-Jewish doctrine of the Church established by Paul and Thomas Aquinas."[206] (Aquinas pictured Jews as "slaves" and "enemies," blasphemers and sinners whose possessions were owned by the Church and princes.[207] Aquinas also wrote about the Jews that "Those who blasphemed against the Son of Man . . . had no excuse, no diminution of their punishment. And so, according to Chrysostom's expositions, the Jews were forgiven their sin neither now nor in the hereafter, for they were punished in this world through the Romans, and in the life to come in the pains of Hell.")[208]

Hitler's Racism

Just as an "umbilical relationship" exists between Old and New Testaments, so is it present in the relationship between Christian and Nazi antisemitism. Hitler's biological racism conflicted with Christian antisemitism but always presupposed it.[209] A purely secular, unChristian racism would have appealed only to the few anti-Christian Nazis who needed a nonreligious justification for their antisemitism.[210] Simultaneously rejecting the Church and imitating its Judaeophobia,[211] Hitler eliminated the Church's more complicated beliefs that the Jews' worst crime in a series of outrages was deicide; that a remnant of Jews would be left to convert at the end of time; and that Jews should disappear through conversion to Christianity. But he agreed with the Church's triumphalistic position that Jews were an alien anti-people who must be eliminated, one way or another. Both Hitler and the churches regarded the Jews as their rivals for men's souls, as scapegoats for historical realities that violated their worldviews, and as enemies to be defeated in proof of the victors' metaphysical power over "evil" and political power over life.[212] Hitler's racism mirrored the great Christian thinkers' triumphalistic antagonism toward the Jews as outlined in the Introduction to this book. Like them, focusing on the Jews' inherent evil and identifying them with the devil, Hitler argued that the Jew "stops at nothing, and in his vileness he becomes . . . the personification of the devil[;] the symbol of all evil assumes the living shape of the Jew."[213] Like traditional Christian anti-semites, Hitler regarded Judaism as the "root of all evil."[214]

At other times, Hitler spoke in biological-racial terms about the Jews. Many writers have focused on this aspect of his ideology. In 1919, Hitler wrote that "the Jews are unquestionably a race, not a religious community."[215] In *Mein Kampf*, he wrote that "race . . . does not lie in the language, but exclusively in the blood . . . ; [the Jew] poisons the blood of others, but preserves his own"; "the lost purity of the blood alone destroys inner happiness forever, plunges man into the abyss for all time, and the consequences can never more be eliminated from body and spirit."[216] In 1942, his table talk included the observation that the way to free Germany of disease was to "dispose of the Jew . . . the racial germ that corrupts the mixture of the blood."[217]

But Hitler admitted a few months before the end of his life that his biological racism had been a sham. Instead, he believed in "spiritual" racism. Many earlier racists paralleled Hitler in regarding spirit not biology as the issue. Paul De Lagarde had written in 1853 that being German was "not a matter of blood, but of a spiritual state of mind" ["*Gemüt, nicht Geblüt*"].[218] Edouard-Adolphe Drumont also defined race as did Hitler. "A race, that is to say, a collection of individuals who think the same, a totality representing a certain number of feelings, beliefs, hopes, aptitudes, traditions . . . ."[219] Arthur Moeller van den Bruck (1876–1925) wrote of "a race of the spirit."[220] Houston Stewart Chamberlain connected his "spiritual" racism with the writings of Paul.[221] "We must agree with Paul, the apostle, when he says: For he is not a Jew who is one outwardly in the flesh, but [a] person is a Jew who is one inwardly, and real circumcision is a matter of the heart—it is spiritual and not literal." Chamberlain's very unracist conclusion was that the exceptional Jew could overcome his Jewish spirit to become a real German, "a wholly humanized Jew is no longer a Jew."[222] German political economist Werner Sombart called for a freeing of all German culture from the corruptions of "the Jewish spirit," as did the German economist and political theorist Eugen Düring and many Nazis, such as Alfred Rosenberg, who argued that it was Judaism that had corrupted Christianity.[223] Xavier Vallat, conservative Catholic and Vichy high commissioner for Jewish Affairs believed that the Jew was always alien not because of any biological "race," but because of his religion. The Jew "is dangerous," wrote Vallat, "not because he is a Semite, but because he is impregnated by the Talmud." Formulating a position for the 1930 elections, the early Nazi leader Gregor Strasser demanded a "German leadership without Jewish spirit." The Conservative Edgar Jung, writing in the 1930s, argued that the alleged racial dichotomy between Aryans and Semites was not the real issue, which was an opposition of Volk against Volk, spirit against spirit. A few days before Hitler came to power, the bishop of Linz observed that Aryans and Christians together had to fight the "dangers and damages arising out of the Jewish spirit."[224] German scientists and writers argued that even should all Jews be killed, their spirit would live on and they would have to continue to fight it.[225]

Likewise, Hitler hated metaphysical Jewishness; he objected to the "Jewish mind" and to the Jewish religious and cultural values that permeated Jewish thought and behavior.[226] Far from being a prophet of race, Hitler did not see the Jews as flesh and blood. Jews were mere symbols of the Jewish spirit he despised. Hitler suggested that beneath the surface of biological racism, a more essential "spiritual racism" existed: individual Jews embodied an inherently evil Jewish spirit—an idea as old as Christian triumphalism.[227] In 1939, Hitler noted "I know perfectly well . . . that in the scientific sense there is no such thing as [a biological] race."[228] A few months before his death, Hitler again differentiated between a race of the mind and one of the body. He told his last private secretary, close associate, and second-in-command, Martin Bormann, that

> we use the term Jewish race as a matter of convenience, for in reality and from the genetic point of view there is no such thing as the Jewish race. There does, however, exist a community. . . . It is [a] spiritually homogeneous group [to] which all Jews throughout the world deliberately adhere . . . and it is this group of human beings to which we give the title Jewish race.[229]

Denying that the Jews were only "a religious entity," since Jewish atheists existed, Hitler described the Jews as "an abstract race of the mind [that] has its origins, admittedly, in the Hebrew religion . . . . A race of the mind is something more solid, more durable that just a [biological] race, pure and simple."[230]

The Final Solution, a Modern Crusade

Every historical event is unique. The Nazi Final Solution to the Jewish Question displays characteristics missing from previous mass-murder campaigns against the Jews. The Third Reich was completely devoted to achieving its murderous ends at all costs, even self-destruction.[231] Although the Nazis often acted opportunistically, they nevertheless formulated a master plan for the mass murder of the Jews throughout Europe. During most of

Jewish history, Jews under attack in one area found a temporary haven in another location, but during the Holocaust, Jews found very few sanctuaries. Indeed, the Allies fighting the Nazis often displayed a callous indifference to the "Final Solution," at other times acted as if the mass destruction of the Jews was to their benefit.[232]

Although the Holocaust was unique in these ways, commonly made distinctions between the Holocaust and previous deadly assaults on the Jews do not bear up under close examination. First, it has often been argued that the modern churches' weakened authority reduced the efficacy of traditional religious restraints on anti-Jewish violence. But historically the Church has often failed to protect Jews from expulsion and mass murder. Although Christian theology and culture provided some protection for Jews, an ideologically powerful and politically tempting Christian anti-Jewishness often stimulated and sustained attacks on Jews.

A second commonly held distinction between Nazi behavior and that of previous Christian actions against the Jews was that the Nazis intended to murder all the Jews, even those loyal to Germany and converted to Christianity, whereas medieval Christians spared those Jews who allowed themselves to be coerced into baptism. At times Christians regarded converted Jews as dangerous, as in fifteenth- and sixteenth-century Spain (and even years later), and devised schemes of legal and social discrimination. But many Christians were also skeptical about the efficacy of baptizing Jews and often did not offer the Jews baptism. Indeed, in Christian societies that included large numbers of converted Jews, the less controllable and the more dangerous Christians believed Jews to be, and schemes were devised to discriminate against them legally and socially. This was as true in fifteenth–seventeenth century Spain as in Germany's Third Reich.

A third false distinction between the Nazi Holocaust and previous Jewish experience within Christendom is the idea that Nazi anti-Jewish ideology, laced with poisonous racism, was *qualitatively* different from Christian anti-Judaism, which was based to a large extent on the accusation of Jewish deicide against Jesus Christ. But Christian racism has existed for centuries, and the Christian accusation that underlies the deicide charge is that Jews were inherently evil long before their attack on Jesus.

Christian antisemitism alone did not cause the Holocaust; but when Christian antisemitism combined with Nazi ideology and modern technology, the resulting "perfect storm" of organized hate made the Holocaust predictable, if not inevitable. Not a necessary consequence of Christianity,[233] the Holocaust was, nevertheless, in a momentous way, the result of the impact that centuries of Christian triumphalism (theology of glory) had on Christians and anti-ecclesiastical Christians. "Products of a culture deeply impregnated with Christian symbols,"[234] the Nazis and their collaborators sought to murder the Jewish spirit whose malevolence the churches had preached for centuries. They had to annihilate Jewish culture, Jewish religion, Jewish Talmuds, Jewish synagogues, and Jewish bodies. They had to destroy Jewishness in many ways like medieval Jew-killers. Granted, in their war against the Jews, Hitler, the Nazis, and the Third Reich supplied an unprecedented organization and mass-murder technology. But the Hebrew and Christian Crusade chronicles convince the reader that—even though Crusader massacres did not represent a systematic attempt by institutional powers to exterminate all Jews—given the hesitant and often nonexistent protection of the Church and Christian princes, if the Crusaders possessed the same organization and technology of death as the Nazis, they would have achieved a final solution of the European Jewish problem a thousand years earlier than the Nazis.[235]

Like the Crusaders and other medieval Christians who slaughtered Jews, Hitler went right to the heart of radical Christian antisemitism: Disregarding the niceties of Christianity's theological distinctions (Augustine's doctrine of the Witness People) as did many a Crusader, he recognized the patristic, medieval, papal, and Lutheran assaults on Jewishness as, in the end, an invitation to murder. In addition to Hitler and the Third Reich, the political, economic, and social crises of the twentieth century, the anti-Jewish climate of opinion, the ideological and emotional bimillennial groundwork, the administrative procedures, the tacit acceptance of mass murder by the vast majority of Christians and the greatest authorities of the time—the social and political elites, the governments, and the churches—caused the Holocaust.[236] These were the sufficient causes needed to produce the Holocaust. This book has attempted to make the case that radical Christian antisemitism—with its deadly combination of theological, racial, ideological, emotional, and behavioral antisemitism—was the most prominent necessary cause, the *sine qua non*, the "that without which," the Holocaust could not have occurred.

The actions of Adolf Hitler were required to achieve the Holocaust the way it happened, although his personal idiosyncrasies were not alone responsible for the centrality of antisemitism in Nazi Germany.[237] Like Voltaire and Wagner, Hitler strongly opposed the churches, but adopted their antisemitism. Like his fierce medieval predecessors, Hitler brushed off traditional limitations on anti-Jewish hostility and transformed preexisting anti-Jewish Christian ideology into action. He acted out an ideology of hatred that had been implicit in Christian theological antisemitism for centuries. Hitler brought "to the surface what has been present before as an internal and broad condition."[238] Like the Crusaders and other medieval Christians who slaughtered Jews by the thousands, Hitler recognized the

Church's hostile, half-hearted, and hypocritical "protection" of Jews as an invitation to slaughter. The only way for Hitler to rid the world of the Jewish spirit was to carry out the eighteenth- to early-nineteenth-century German nationalist Johann Fichte's suggestion and to "cut all their Jewish heads off." Hitler and his collaborators compacted the worst of the previous 1,900 years of crimes against Jews into one 12-year assault.[239] Not without a crucial streak of opportunism, Hitler was an ideologue who believed in his ideas and was totally devoted to carrying them out. He articulated what most Germans already felt about the Jews, and he did so in a riveting, charismatic, attractive, seductive manner. A demagogue, yes, but Hitler did not persuade an unwilling audience; he confirmed his audience's preexisting beliefs and feelings.[240] Like his fierce medieval predecessors, Hitler brushed off traditional limitations on anti-Jewish hostility. He acted out an ideology of hatred that had been implicit in Christian religious and racial anti-semitism for centuries. Tamas Nyiri, an observer of the Holocaust in Hungary, has written, "The Holocaust is no theological accident . . . . The anti-Judaism of the Middle Ages is shockingly close to Hitlerian racism."[241] Contemporary crises—which may have molded the specific form of the war against the Jews—brought traditional anti-Jewish attitudes to consciousness, but they did not cause them.

After the Nazi seizure of power, Hubert Lanzinger painted Hitler mounted on a horse, fitted out in armor, holding a Nazi *flag—Der Bannerträger* ("The Standard Bearer")—Hitler's official portrait as a medieval knight on Crusade.[242] In 1938, it was reported that Hitler admired a poem of Dietrich Eckart's that proclaimed: "Father in Heaven, resolved to the death we kneel before you . . . . Does any other people more loyally follow your awful command than we Germans do . . . ? Up and onward, onward to the holy Crusade."[243] Leading a great and powerful nation, Hitler could bring resources to bear against the Jews that the early Crusader leader Emicho of Leiningen could only dream of. Hitler created a system in which the Jews' souls were crushed, their religion lost, their morality murdered, and their bodies, like their dreams, transmuted into smoke and ashes. He catalyzed the centuries-old brooding anti-Jewish antipathy of the great mass of the Christian populace, from the ordinary layperson to the highest-ranking Christian secular and religious officials, into terrifying actuality.

Postscript

Unlike the Middle Ages, when some popes acted "as if they were God himself"—according to Pope Innocent III, "Every cleric must obey the pope, even if he commands what is evil; for no one may judge the pope"[1]—modern popes' attitudes, policies, and actions set the moral tone, did not determine the behavior, of the Church. Yet until 1965 (the year of the Second Vatican Council's *Nostra Aetate*, which declared that all Jews were not responsible for Christ's death and that antisemitism was a sin), popes continued to follow the 1500-year tradition of Augustine's "Witness People." The devilish deicide Jews were the world's leading evildoers, and should suffer but not be killed.

Not until Vatican Council II did the Roman Catholic Church commence a reconciliation with the Jewish people. In 1965, the Council asserted the Jews' common spiritual heritage with Catholics, the Jews' lack of collective guilt for the death of Christ, Israel's freedom from God' curse, and the special loving relationship between Catholics and Jews.[2] It condemned all examples of antisemitism.[3] But *Nostra Aetate* did not ask forgiveness from the Jews for the Church's past antisemitism nor did it assert the contemporary validity of Judaism.

Over the past 40 years, the Church has published Guidelines condemning antisemitism. In 1986 John Paul II became the first pope in modern times to visit a synagogue, and he apologized for Catholic antisemitism. In 1998, the Vatican published *We Remember: A Reflection on the Shoah*. It encouraged Catholics everywhere to repent of past sins and examined the fact that the Holocaust occurred in Christian Europe.

At the same time, John Paul II canonized Pius IX and attempted to do the same for Pius XII—both over Jewish objections.

Despite these setbacks, John Paul II regarded the Jews as "the dearly beloved elder brother of the Church." Seven years after the 1993 Vatican establishment of diplomatic relations with Israel, John Paul II visited the Holocaust memorial in Jerusalem and the Wailing Wall. As a result, the pope's condemnation of antisemitism as "a sin against God and man" became more widely accepted among Catholics.

Even if the effect of the Church's transformation of its relations with Jews has not yet fully reached the whole Church, its potential for improving Jewish-Catholic relations is unlimited.

Notes

Chapter 1 Christianity, Antisemitism, and the Holocaust

Notes to pages 2–5

1. Martin Luther, "Heidelberg Disputation," Article 21, in *Luther's Works*, 31:40.
2. Jacob Neusner, "Christian Missionaries—Jewish Scholars," *Midstream* (October 1991), 31.
3. Cited in Richard Steigmann-Gall, *The Holy Reich: Nazi Conceptions of Christianity, 1919–1945* (Cambridge 2003), 1, 4.
4. Peter Hayes, ed., *Lessons and Legacies* (Evanston 1991), 8.
5. Raul Hilberg, *The Destruction of the European Jews* (Chicago 1967), 5–7.
6. See Joshua Trachtenberg, *The Devil and the Jews* (Philadelphia 1961).
7. See Pierre Quillard, *Le Monument Henry* (Paris 1899); also Stephen Wilson, "Le Monument Henry: La structure de l'antisémtisme en France, 1898–1899," *Annales* 32 (1977).
8. Steigmann-Gall, *The Holy Reich*, 112. The only paganist to reject Christ was General Ludendorff, and he was the only paganist expelled from the Nazi Party.
9. Steigmann-Gall, *The Holy Reich*, 3; summarized, 10–12.
10. Nuremberg War Crimes Trial Document 1708-PS.
11. Nuremberg War Crimes Trial Document 1708-PS, 49–50.
12. *Mythus des 20. Jahrhunderts* (Munich 1930), 391.
13. See Klaus Fischer, *The History of an Obsession: German Judaeophobia and the Holocaust* (New York 1998).
14. Marcel Simon, *Verus Israel* (Oxford, Eng. 1986), 398.
15. Robert Alter, "From Myth to Murder," *The New Republic* (20 May 1991), 34, 37–8; Walter Sokel, "Dualistic Thinking and the Rise of Ontological Antisemitism in 19th-Century Germany," in Sander Gilman and Steven Katz, eds., *Antisemitism in Times of Crisis* (New York 1991), 154–72.
16. Quoted in Bernard Wasserstein, *Britain and the Jews of Europe, 1939–1945* (London 1979), 47.
17. Gerhart Ladner, "Aspects of Patristic Anti-Judaism," *Viator: Medieval and Renaissance Studies* 2 (1971), 362.
18. Augustine, *Adversus Judaeos* 7, 10, see also 8, 11. "*Occidistis Christum in parentibus vestris.*"

19. Jerome, *The Homilies of Saint Jerome* (Washington, DC 1964), 1:255, 258–62.
20. John Chrysostom, *Homilies against Judaizing Christians*, 6.2.10.
21. *Contra Judaeos*, 1, 18, in Rosemary Ruether, *Faith and Fratricide* (New York 1965), 130.
22. Quoted by James Parkes, *The Conflict of Church and Synagogue* (New York 1979), 290.
23. "On the Sabbath," 4:23, in Ruether, *Faith and Fratricide*, 148.
24. Mary Stroll, *The Jewish Pope: Ideology and Politics in the Papal Schism of 1130* (Leiden 1987), 166.
25. *Summa Theologiae*, 1a 2ae, 102, 6.8.
26. See Yosef Yerushalmi, *Assimilation and Racial Antisemitism* (New York 1982); Léon Poliakov, *The Aryan Myth* (New York 1974).
27. Poliakov, *The Aryan Myth*, 12–13; Albert Sicroff, *Les controverses des statuts de "pureté de sang" en Espange du XVe au XVIIe siècle* (Paris 1960); and Michael Glatzer, "Pablo de Santa Maria on the Events of 1391," in Shmuel Almog, ed., *Antisemitism through the Ages* (Oxford 1988), 127–37.
28. Cecil Roth, *A History of the Marranos* (New York 1974), 21, 29–30.
29. Yerushalmi, *Assimilation and Racial Antisemitism*, 10.
30. Yerushalmi, *Assimilation and Racial Antisemitism*, 9.
31. Sandoval, *Historia de la vida y hechos del emperator Carlos V*, quoted by Jerome Friedman, "Jewish Conversion, the Spanish Pure Blood Laws and Reformation: A Revisionist View of Racial and Religious Anti-Semitism" (*Sixteenth-Century Journal*), 18 (1) (1987), 16–17.
32. Vincente da Costa Mattos, *Breve discurso contra a heretica perfidia do judaismo*, quoted by Friedman, "Jewish Conversion, the Spanish Pure Blood Laws and Reformation."
33. Torrejoncillo, *Centinela contra Judios puesta en la torre de la iglesia de Dios* (published in 1673, expanded in 1736). See Frank Manuel, *Broken Shaft* (Cambridge, MA 1992), 223–4, and Yerushalmi, *Assimilation and Racial Antisemitism*, 16.
34. James Reites, S.J., "St. Ignatius of Loyola and the Jews," *Studies in the Spirituality of Jesuits* (September 1981), 15, 16, 32.
35. Reites, "St. Ignatius of Loyola and the Jews."
36. Peter Godman, *Hitler and the Vatican* (New York 2004), 62.
37. Shlomo Simonsohn, *The Apostolic See and the Jews: Documents, 1464–1521* (Toronto 1990), docs. 879, 1157, 1158, 1167, 1206, 1334 (hereafter cited as Simonsohn I, II, III); and Shlomo Simonsohn, *The Apostolic See and the Jews: History* (Toronto 1991), 385, 387–91 (hereafter cited as Simonsohn, *History*).
38. Simonsohn III, docs. 1167 and 1206.
39. Owen Chadwick, *The Popes and European Revolution* (Oxford 1981), 133.
40. To Pirkheimer and Reuchlin in November 1517. See Heinrich Graetz, *The History of the Jews* (Philadelphia 1940), 4:435, 443–4; Simonsohn, *History*, 333 n101; Heiko Oberman, *The Roots of Antisemitism in the Age of Renaissance and Reformation* (Philadelphia 1981), 30–1, 53.
41. Quoted by Léon Poliakov, *The History of Antisemitism* (New York 1974), 1:215.

42. Quoted by Sander Gilman, "Martin Luther and the Self-Hating Jews," in Gerhard Dünnhaupt, ed., *The Martin Luther Quincentennial* (Detroit 1985), 84–8.
43. Quoted in Abraham Geiger, ed., *Johann Reuchlin: Sein Leben und Seine Werke* (Leipzig 1871); and Allan Gould, *What Did They Think of the Jews?* (New York 1991), 49.
44. See Graetz, *The History of the Jews*, 4:435, 443–4; Simonsohn, *History*, 333 n101; Oberman, *The Roots of Antisemitism*, 30–1, 53.
45. Martin Luther, "That Jesus Christ Was Born a Jew," in Walther Brandt, ed., *Luther's Works* (Philadelphia 1967), 45:213; repeated in *Vom Schem Hamphoras*.
46. Martin Luther, "On the Jews and Their Lies," in *Luther's Works*, trans. Franklin Sherman (Philadelphia 1971), 137–8 (this translation of "Von den Juden und Ihren Lügen," Weimar Edition, 53:417–52, will hereafter be abbreviated as "Jews").
47. Reinhold Lewin, *Luthers Stellung zu den Juden: Ein Betrag zur Geschichte der Juden während des Reformationszeitalters* (Berlin 1911), 77.
48. Luther, "Jews," 170, 216–17, 253, 267–9, 285–6. See also Gerhard Falk, *The Jew in Christian Theology* (London 1992), 166–7.
49. Yerushalmi, *Assimilation and Racial Antisemitism*, 17, 21.
50. Quoted in David Kertzer, *The Popes against the Jews: The Vatican's Role in the Rise of Modern Anti-Semitism* (New York 2001), 136–8.
51. Quoted in Kertzer, *The Popes against the Jews*, 146.
52. Kertzer, *The Popes against the Jews*, 208.
53. Kertzer, *The Popes against the Jews*, 210–11.
54. Charles Glock and Rodney Stark, *Christian Beliefs and Antisemitism* (New York 1966), xvi, 185–7, 50–65, 73–4, 105. See also Rodney Stark et al., *Wayward Shepherds* (New York 1971), 5, 9–10, 50; Alphons Silbermann, *Sind Wir Antisemiten?* (Cologne 1982), 51–2.
55. Gordon Allport, *The Nature of Prejudice* (Garden City 1958), 446.
56. Frank Felsenstein, *Antisemitic Stereotypes: A Paradigm of Otherness in English Popular Culture, 1660–1830* (Baltimore 1995); Quillard, *Le Monument Henry*.
57. Mahzarin Banaji and Anthony Greenwald using the Implicit Association Test. See Sharon Begley, "The Roots of Hatred," *The AARP Magazine* (May–June 2004), 48.
58. Michael Shermer, *The Science of Good and Evil*, quoted by Sharon Begley, "The Roots of Hatred."
59. Begley, "The Roots of Hatred," 48.
60. Jakob and Wilhelm Grimm, *Deutsches Wörterbuch*, 2nd ed. (Leipzig 1877), Vol. 4, S. 2353.
61. Célia Szniter Mentlik, "HISTÓRIA, LINGUAGEM E PRECONCEITO: ressonâncias do período inquisitorial sobre o mundo contemporâneo," *Revista História Hoje* 2 (5) (November 2004); http://www.anpuh.uepg.br/historia-hoje/vo12n5/celia.htm
62. *Oxford English Dictionary* (Oxford 1933, 1961), 5:576–7; and the *Oxford English Dictionary Supplement* (Oxford 1976), 2:18–19.
63. See, for example, Fischer, *The History of an Obsession*.

64. Kertzer, *Popes against the Jews*, 206.
65. Based on Allport's list in his *Nature of Prejudice*, 14–15.
66. Gen. 4:9 and Deut. 19:10; see also Num. 35:9–31 and Josh. 20:1–9. See also Phillip Hallie, *Lest Innocent Blood Be Shed* (New York 1980).
67. Karl Jaspers, *The Question of German Guilt* (New York 1947), 32.

CHAPTER 2 THE CHURCH FATHERS

1. Moshe Herr, "The Sages' Reaction to Antisemitism in the Hellenistic World," in Almog, *Antisemitism through the Ages*, 27.
2. Molly Whittaker, *Jews and Christians: Graeco-Roman Views* (Cambridge, U.K. 1984), 190, 148; Robert Wilken, "The Christians as the Romans (and Greeks) Saw Them," in E.P. Sanders, ed., *Jewish and Christian Self-Definition* (Philadelphia 1980), 1:105, 120.
3. Menachem Stern, *Greek and Latin Authors on Jews and Judaism* (Jerusalem 1980).
4. Wilken, "The Christians as the Romans (and Greeks) Saw Them," 62–4.
5. H.H. Ben-Sasson, "Effects of Religious Animosity on the Jews," in H.H. Ben-Sasson, ed., *A History of the Jewish People* (London 1976), 411; Gavin Langmuir, "Anti-Judaism as the Necessary Preparation for Antisemitism," *Viator: Medieval and Renaissance Studies* 2 (1971), 385.
6. See Marcel Simon, *Verus Israel: Etude sur les relations entre chrétiens et juifs dans l'empire romain, 135–425* (Paris 1964); Jules Isaac, *Jesus and Israel* (New York 1971); Jean Juster, *Les Juifs dans l'Empire romain* (Paris 1914), 1:45–8 n1.
7. Pierre Pierrard, *Juifs et Catholiques Français* (Paris 1970), 298.
8. Norman Beck, *Mature Christianity* (London 1985), 223.
9. Rosemary Ruether, "Christology and Jewish–Christian Relations," in Abraham Peck, ed., *Jews and Christians after the Holocaust* (Philadelphia 1982), 27, 34–5.
10. David Efroymson, *Tertullian's Anti-Judaism and Its Role in His Theology* (Philadelphia, Temple University Dissertation, 1975), 1, 4, 14, 22, 82, 85, 109; Nicholas De Lange, *Origen and the Jews* (Cambridge, U.K. 1976), 78–83.
11. Wilken, "Insignissima Religio, Certe Licita? Christianity and Judaism in the Fourth and Fifth Centuries," in Jerald Brauer, ed., *The Impact of the Church Upon Its Culture* (Chicago 1968), 49; John Gager, *The Origins of Anti-Semitism* (New York 1983), 157.
12. Gager, *The Origins of Anti-Semitism*, 260.
13. David Flusser, "Foreword," to Clemens Thoma, *A Christian Theology of Judaism* (New York 1980), 7.
14. Wilhelm Bousset, *The Antichrist Legend: A Chapter in Christian and Jewish Folklore* (London 1896), 158, 166–70.
15. Robert Wilken, *Judaism and the Early Christian Mind: A Study of Cyril of Alexandria's Exegesis and Theology* (New Haven 1971), 174, 226–8. See also Peter Richardson, ed., *Anti-Judaism in Early Christianity: Paul and the Gospels* (Waterloo, Canada 1986).

16. For definitions, see F.L. Cross, ed., *The Oxford Dictionary of the Christian Church* (Oxford 1983). For summaries of their works, see Heinz Schreckenberg, ed., *Die christlichen Adversus–Judaeos–Texte und ihr literarisches und historisches Umfeld, 1.–11. Jh.* (Frankfurt 1982); and A.L. Williams, ed., *Adversus Judaeos: A Bird's Eye View of Christian Apologiae Until the Renaissance* (London 1935).
17. Benard Lewis, *The Middle East: A Brief History of the Last 2000 Years* (New York 1995), 45.
18. David Rokeah, "The Church Fathers in Writings Designed for Internal and External Use," in Almog, *Antisemitism through the Ages*, 57.
19. Simon, *Verus Israel* (Eng.), 70.
20. Jaroslav Pelikan, *The Spirit of Eastern Christendom* (Chicago 1974), 211.
21. Gager, *The Origins of Antisemitism*, 164.
22. Isidore Loeb, "La Controverse religieuse entre les Chrétiens et les Juifs au Moyen Age," *Revue de l'Histoire des Religions* 17 (1888), 318–19.
23. "Tractatus in LI Psalmum," 6; J.-P. Migne, ed., *Patrologiae, Cursus Completus, Series Latina* (Paris 1844-) (unless otherwise indicated, translations were executed by the author).
24. Robert Markus, *The End of Ancient Christianity* (Cambridge, Eng. 1990), 1.
25. See Ben Zion Bokser, *Judaism and the Christian Predicament* (New York 1967), 35; Gavin Langmuir, *History, Religion, and Antisemitism*, (Berkeley 1990), chapter 14; Simon, *Verus Israel* (Eng.), 96.
26. Jerome to Augustine, Epistle 75, Augustine to Jerome, Epistle 82; in Simon, *Verus Israel* (Eng.), 93–4.
27. Chrysostom, *Homilies against Judaizing Christians*, 1.6.5, 1.6.8.
28. Wilken, "Insignissima Religio, Certe Licita?" 51.
29. Stephen G. Wilson, "Marcion and the Jews," in Stephen G. Wilson, ed., *Anti-Judaism in Early Christianity* (Waterloo, Canada 1986), 2:58.
30. Exod. 19:6.
31. In *Early Christian Writings* (Harmondsworth, Eng. 1968), 197.
32. See Rom. 2:25–9; Gal. 5.
33. Rom. 9:8–13.
34. Juster, *Les Juifs dans l'Empire romain*, 1:264 n11.
35. Parkes, *Church and Synagogue*, 83–4, 104, 278.
36. Gen. 22; Deut. 12:15–16, 23–5, 18:10; Lev. 3:17, 7:26, 17:10–14; Jer. 2:35, 19:4–6.
37. Gen. 4:8–15.
38. Matt. 23:31–6.
39. Abraham, Jacob, and Moses were three examples. Gen. 13; Gen. 32; Exod. 32. God called Jacob "Israel."
40. Gen. 32:28.
41. *Homily against the Jews*, I, 283.
42. In Cyril of Alexandria's comment on Article 8 of the Apostles' Creed. Juster, *Les Juifs dans l'Empire romain*, 1:302.
43. In both the Torah and Talmud: for example, Lev. 11:7–8; Menahoth 64b.
44. "On the Sabbath against the Jews," in *De Fide Orthodoxa*, 4:23, in Ruether, *Faith and Fratricide*, 128.
45. Andrew Sharf, *Byzantine Jewry: From Justinian to the Fourth Crusade* (New York 1971), 27; Juster, *Les Juifs dans l'Empire romain*, 1:231 n8.

46. Wilken, "Insignissima Religio, Certe Licita?" 51.
47. Chrysostom, *Homilies against Judaizing Christians*, 1.7.5, 2.1.1, 2.3.4. John Chrysostom, *Saint John Chrysostom: Discourses against Judaizing Christians*, translated with Introduction by Paul Harkins (Washington, DC 1979). All translations from these homilies are taken from the Paul Harkins edition.
48. Chrysostom, *Homilies against Judaizing Christians*, 1:3:1.
49. Quoted by Friedrich Heer, *God's First Love* (New York 1967), 37.
50. Ambrose, "Epistolarum Classis I. XL. 26," *PL*, 15:1110–11. Reprinted in Jacob Marcus, *The Jew in the Medieval World: A Source Book, 315–1791* (New York 1979), doc. 21:i.
51. The theme was also proclaimed in the *Didascalia*, which damned the Jews to hell but noted that Christians were obliged to call Jews their brothers. *Didascalia*, 5:14:23 and 5:20.
52. Rom. 11:28; Gal. 2:16, 3:7; Eusebius, *Ecclesiastical History*, 1.4.6; Augustine, *City of God*, 15.1, and *Against Two Epistles of the Pelagians*, 3.4.11; Justin Martyr, *Dialogue with Trypho*, 119, 3–6. Aphraates cites Paul, Rom. 11:17.
53. *Dialogue with Trypho*, 11:5.
54. John. 8:39, 44. Aphraates quoted in F.E. Talmage, ed., *Disputation and Dialogue* (New York 1975), 27; Jeffrey Siker, *Disinheriting the Jews: Abraham in Early Christian Controversy* (Louisville 1991), 193.
55. Augustine, *Tractatus Adversus Judaeos*, I–II.
56. Augustine, *Tractatus Adversus Judaeos*, III, iii.
57. Augustine, "Contra Adversarium Legis et Prophetarum," *PL*, 42:623
58. Justin, *Apologia* 1:53, and Origen, *Contra Celsum* 7:26.
59. Barnabas, Justin, and Augustine quoted in Simon, *Verus Israel* (Eng.), 71, 147.
60. Irenaeus, *Contra Haereses*, III, 21.
61. Ruether, *Faith and Fratricide*, 147.
62. Cross, *The Oxford Dictionary of the Christian Church*, 1352.
63. Robert Wilde, *The Treatment of the Jews in the Greek Christian Writers of the First Three Centuries* (Washington 1949), 149.
64. Lellia Cracco Ruggini, "Pagani, Ebrei, e Christiani: Odio Sociologico e Odio Teologico nel Mondo Antico," in *Gli Ebrei nell'Alto Medioevo* (Spoleto 1980), 46; Amos Funkenstein, "Basic Types of Christian Anti-Jewish Polemic in the Later Middle Ages," *Viator* 2 (1971), 374. For a collection, see Williams, *Adversus Judaeos.*
65. Efroymson, *Tertullian's Anti-Judaism*, 1, 4, 11, 64, 82, 85, 109.
66. Quoted by Efroymson, *Tertullian's Anti-Judaism*, 125.
67. Jerome, "Epistolae LXXXXIII and CXXI," in J.-P. Migne, ed., *Patrologiae, Cursus Completus, Series Latina*, 22:699, 22:1032 (unless otherwise indicated, translations were executed by the author).
68. Jerome, *De Antichristo in Danielem* 4, 11:21–30, in F. Glorie, ed., *Commentarium in Danielem, Libri 3–4* [*Corpus Christinaorum, Series Latina*] (Turnhout 1964), 75A:917–20. Isidore of Seville followed the same path. See B.-S. Albert, "Isidore of Seville: His Attitude towards Judaism and His Impact on Early Medieval Canon Law," *The Jewish Quarterly Review* (January–April 1990), 212.
69. Rom. 2:29.
70. Rom. 7:5–6.

71. Rom. 8:1–11.
72. 1 Cor. 2:11.
73. *Early Christian Writings*, 207.
74. Mark. 14:10; Luke 22:5; John 13:2; Matt 26:14–15, 27:3, alluding to Zech. 11:12.
75. Elisabeth Revel-Neher, *The Image of the Jew in Byzantine Art* (Oxford 1992), 22.
76. "*Specialiter intelligitur de Juda: generaliter autem de Judaeis,*" *PL*, 26:1224.
77. Jerome, *The Homilies of Saint Jerome*, 1:255, 258–62.
78. My people, what have I done to you? . . . You prepared a cross for your savior." p. 193.
79. Jerome also argued that forgiveness should be given since God had not destroyed the Jews and because Judaism was the root of Christianity. Jerome, *The Homilies of Saint Jerome*, 263–4, 255–7, 267.
80. Quoted in Kertzer, *Popes against the Jews*, 212.
81. Jacques Nobécourt, "*Le Vicaire" et l'Histoire*, (Paris 1964), 207–8.
82. Simon, *Verus Israel* (Eng.), 231.
83. "*The Anguish of the Jews*: A Theology of Intolerance," *Continuum* (Autumn 1966), 419.
84. Letter 112 to Augustine, quoted by Terrance Callan, *Forgetting the Root* (New York 1986), 88.
85. Chadwick, *The Popes and European Revolution*, 20–1.
86. See "Réflexions sur la genèse du discours antisémite," in Valentin Nikiprowetzky, ed., *De l'antijudaïsme antique à l'antisémitisme contemporaine* (Lille 1979), 281; Bernard Glassman, *Antisemitism without Jews* (Detroit 1975).
87. See Gavin Langmuir, "Toward a Definition of Antisemitism," in Helen Fein, ed., *The Persisting Question: Sociological Perspectives and Social Contexts of Modern Antisemitism* (Berlin 1987), 127.
88. Poliakov, *The History of Antisemitism*, 3:288.
89. Augustine, *Tractatus Adversus Judaeos*, 15.
90. Augustine, *Tractatus Adversus Judaeos*, 7:9, in *PL*, 42:57.
91. Augustine's Commentary on Psalm 75:1 (*PL*, 36:958).
92. Bernhard Blumenkranz, "Augustin et les juifs: Augustin et le judaïsm," in Blumenkranz, ed., *Juifs et chrétiens: Patristic et Moyen Age* (London 1977), 226, 230–1, 235–7.
93. Jeffrey Russell, *The Devil: Perceptions of Evil from Antiquity to Primitive Christianity* (Ithaca 1977), 240 n25.
94. Ambrose, *Cain and Abel*, in Ruth Mellinkoff, *The Mark of Cain* (Berkeley 1981), 92–3. See also Langmuir, "Toward a Definition of Antisemitism," 74.
95. In his Commentary on Psalm 58.
96. Augustine, "Commentaries on Psalms 58 and 59," in *PL*, 36–7:705.
97. Augustine, "Reply to Faustus, the Manichaean," in Talmage, *Disputation and Dialogue*, 31. See also Augustine, *City of God* (New York 1950), 656–8, Book 18, chapter 46; Thomas Merton, "Preface," *The City of God* (New York 1950), xi.
98. "Augustin et les juifs: Augustin et le judaïsm," 231. He repeated his doctrine at least 20 times in his work. Juster, *Les Juifs dans l'Empire romain*, 1:228 n6. See also Bernhard Blumenkranz, *Die Judenpredigt Augustins* (Basle 1946).

99. Amnon Linder, ed., *The Jews in Roman Imperial Legislation* (Detroit 1987), 241n7.
100. Augustine, "Reply to Faustus, the Manichaean," in Talmage, *Disputation and Dialogue*, 31.
101. See Mellinkoff, *The Mark of Cain*, 38–40.
102. Ambrose, "Epistolarum Classis I, Epistola XL," *PL*, 16:1105, 1109.
103. Severus Majoricensis, "Epistola de Judaeis," *PL*, 20:731–3.
104. Linder, *The Jews in Roman Imperial Legislation*, 284, 288.
105. Quoted by Heer, *God's First Love*, 37.
106. Jerome, "Commentariorum in Amos Prophetam," Liber II, Caput V, in *PL*, 25:1054.
107. Juster, *Les Juifs dans l'Empire romain*, 2:207–14.
108. Joseph Klausner, *Jesus of Nazareth* (New York 1925), 47–54; James Parkes, *The Jew in the Medieval Community* (New York 1976), 54.
109. *Liber Sacramentorum Romanae Ecclesiae, Assemani, Cod. Lt.*, IV ii, 91, in Parkes, *Church and Synagogue*, 401.
110. See Ambrose, "Epistolarum Classis I. Epistola XL," in *PL*, 16:1105, 1109.
111. "Interioris mentis sordibus." Ambrose, "Exposito Evangelii Secundum Lucam. Libris X," in *PL*, 15:1630.
112. Thomas Aquinas, *Exposito Continua*, also know as the Golden Chain, *Aurea Catena in quatuor Evangelia, Expositio in Lucam.* "Commentary on the Four Gospels, collected out of the Works of the Fathers. By S. Thomas Aquinas." [85793] Catena in Lc., cap. 4 l. 7.
113. Ambrose, "Epistolarum Classis I. XL. 26," in *PL*, 15:1110–11.
114. Ramsay MacMullen, *Christianizing the Roman Empire* (New Haven 1984), 100.
115. Acts 11:26.
116. Simon, *Verus Israel* (Eng.), 401–2.
117. Marcus, *The Jew in the Medieval World*, doc. 19.
118. Chrysostom, *Homilies against Judaizing Christians*, 1.5.1. See also 1.4.5.
119. Chrysostom, *Homilies against Judaizing Christians*, 4.7.4; see also 1.4.1.
120. See Fred Grissom, *Chrysostom and the Jews* (Southern Baptist Theological Seminary Dissertation 1978), 166.
121. Chrysostom, *Homilies against Judaizing Christians*, 1.6. 8 (my emphasis).
122. Chrysostom, *Homilies against Judaizing Christians*, 6.2.10.
123. Chrysostom, *Homilies against Judaizing Christians*, 1.7.2, 4.3.6.
124. *Apostolic Constitutions*, 2.61.1; Chrysostom, *Homilies against Judaizing Christians*, 1.5.1.
125. Chrysostom, *Homilies against Judaizing Christians*, 5.1.7.
126. John Chrysostom, Homily on Psalm 109, in J.-P. Migne, ed., *Patrologiae, Cursus Completus, Series Graeca* (Paris, 1886–), 55:267; *Saint John Chrysostom: Homilies on Genesis* (Washington DC 1986), Homily 8:6, 1:108.
127. Chrysostom, *Homilies against Judaizing Christians*, 1.5.2; see also 6.6.11.
128. *Saint John Chrysostom: Homilies on Genesis*, Homilies 8:6–7, 39:18–19, 40:17, 1:108, 2:386–7, 397.
129. Chrysostom, *Homilies against Judaizing Christians*, 1.7.1; Chrysostom, Homilies 45 and 84, in *Commentary on Saint John the Apostle and Evangelist* (Washington DC 1969), 1:448–9, 2:419.

130. *Apostolic Constitutions*, 2.61.1; Chrysostom, *Homilies against Judaizing Christians*, 1.3.1, 1.3.5, 1.4.2.
131. Chrysostom, *Homilies against Judaizing Christians*, 1.6.3, 6.7.5–6.
132. Chrysostom, *Homilies against Judaizing Christians*, 1.5.2.
133. Chrysostom, *Homilies against Judaizing Christians*, 1.2.4–6. See also John Chrysostom, *Demonstration to the Jews and Gentiles That Christ Is God*, 4. *PG* 48:819. Sad to say, neither statement was discussed by Robert Wilken in his otherwise fine book, *John Chrysostom and the Jews* (Berkeley 1983).
134. Chrysostom, *Homilies against Judaizing Christians*, 4.1.6.
135. See Heer, *God's First Love*, 130, 284–6; Adolf Hitler, *My New Order*, ed. Raoul de Roussy de Sales (New York 1973), 24.
136. See Wilken, *John Chrysostom and the Jews*, 126 n.
137. Simon Dubnow, *History of the Jews in Russia and Poland* (Philadelphia 1920), 31–2; Wilken, *John Chrysostom and the Jews*, 162.
138. Malcolm Hay, *Thy Brother's Blood* (New York 1975), 27.
139. Juster, *Les Juifs dans l'Empire romain*, 1:262.
140. Sharf, *Byzantine Jewry*, 3.
141. J.F. Haldon, *Byzantium in the Seventh Century* (Cambridge, Eng. 1990), 345. See also Sharf, *Byzantine Jewry*; A.H.M. Jones, *The Later Roman Empire* (Oxford 1964); and Ruggini, "Pagani, Ebrei, e Christiani," 1:13–101.
142. *Codex Justinianus* 1:9:9; *Codex Theodosianus* 16:8:11; *C.T.* 16:8:12; *C.T.* 16:8:13.
143. C.T. 12:1:158; *C.T.* 16:8:14. Walter Pakter, *De His Qui Foris Sunt: The Teachings of the Medieval Canon and Civil Lawyers Concerning the Jews* (Ph.D. Dissertation, Johns Hopkins University, 1974), 323 n2.
144. Parkes, *Church and Synagogue*, 231–2.
145. *C.T.* 16:8:15; *C.T.* 16:8:17.
146. Revel-Neher, *The Image of the Jew in Byzantine Art.*
147. Pelikan, *The Spirit of Eastern Christendom*, 200.
148. Theophranes, in Revel-Neher, *The Image of the Jew in Byzantine Art*, 12–13.
149. Juster, *Les Juifs dans l'Empire romain*, 1:469 n1, 231 n7.
150. Linder, *The Jews in Roman Imperial Legislation*, 47; Parkes, *Church and Synagogue*, 246.
151. Sharf, *Byzantine Jewry*, 20–1.
152. *C.T.* 16:8:2, 16:8:8, 16:8:9, 16:8:13, 16:8:20.
153. *C.J.* 1:5:12–13, 21; 1:3:54; 1:10:2; *Novella* 45. Other laws associating Jews and heretics prevented them from acquiring Catholic property: *C.J.* 1:5:10 and *Novella* 131.
154. *C.J.*, *Novellae* 139, 37.
155. Chrysostom, *Homilies against Judaizing Christians*, 6:6.
156. Published in February 553. *C.J.*, *Novella* 146.
157. Heraklios returned the supposed true cross to Jerusalem in 629. He may have influenced Dagobert I (628–39), the Merovingian king in France, and Catholic Visigothic King Sisebut to anti-Jewish actions. Parkes, *Church and Synagogue*, 265–6; Roth, *A History of the Marranos*, 7.
158. Leo was a religious reformer who forbade Jews to hold public office and punished conversion to Judaism. He published several edicts forbidding the worship of images, beginning the Iconoclastic controversy. Although those on

both sides of the controversy were anti-Jewish, those who opposed the emperor argued that the Jews were responsible for the imperial attacks on images.

159. Steven Bowman, *The Jews of Byzantium, 1204–1453* (Montgomery, AL 1985), 9, 17–18.
160. Bowman, *The Jews of Byzantium*, 10.
161. Bowman, *The Jews of Byzantium*, 9.
162. Revel-Neher, *The Image of the Jew in Byzantine Art*, 13–14.
163. "Responsum of Demetrius Khomagianos, Archbishop of Ochrida," in Bowman, *The Jews of Byzantium*, doc. 18.
164. Benjamin of Tudela, *Sefer ha-Masa'oth*, in Bowman, *The Jews of Byzantium*, 335.
165. Bowman, *The Jews of Byzantium*, 18–20.
166. Benjamin of Tudela, in Revel-Neher, *The Image of the Jew in Byzantine Art*, 17 n48.
167. Bowman, *The Jews of Byzantium*, 32.
168. Gennadios Scholarios, "Refutation of the Jewish Error" (1464), in Bowman, *The Jews of Byzantium*, doc. 145.
169. Bowman, *The Jews of Byzantium*, 37.
170. Bowman, *The Jews of Byzantium*, docs. 1and 2.
171. Bowman, *The Jews of Byzantium*, doc. 33.
172. Bowman, *The Jews of Byzantium*, doc. 35
173. Bowman, *The Jews of Byzantium*, doc. 31.
174. Matthew Blastares, *Syntagma*, in Bowman, *The Jews of Byzantium*, doc. 57.
175. Gregory of Nyssa, "In Christi Resurrectionem, Oratio V," in *PG* 46:685–6.
176. Wilde, *The Treatment of the Jews*, 78–225. See also Maxwell Staniforth, trans., *Early Christian Writings: The Apostolic Fathers* (New York 1981).
177. See Russell, *The Devil*, 240.
178. *Sermon on the Resurrection*, included in the writings of John Chrysostom, *PG*, 61:733.
179. Pierrard, *Juifs et Catholiques Français*, 298.
180. See Langmuir, "Toward a Definition of Antisemitism," 348.
181. Simonsohn, *History*, 292.
182. Jaroslav Pelikan, *The Growth of Medieval Theology* (Chicago 1978), 247; Gavin Langmuir, "*The Anguish of the Jews*: The Enervation of Scholarship," *Continuum* (Autumn 1966), 624.
183. Loeb, "La Controverse religieuse entre les Chrétiens et les Juifs au Moyen Age," 313–14.
184. Justinian's anti-Jewish law of 553 forbade the study of Mishna. *Novella* 146, see Linder, *The Jews in Roman Imperial Legislation*, doc. 66.
185. *Jerusalem Talmud*, Ta'an, 2:1, 65–6; see also *Exodus Rabah*, 29–35.
186. The medieval Church did censor portions of the Talmud considered anti-Christian.
187. See Simon, *Verus Israel* (Eng.), 401–2; Douglas Hare, "The Rejection of the Jews in the Synoptic Gospels and in Acts," in Alan Davies, ed., *Antisemitism and the Foundations of Christianity* (New York 1979), 30–1; Steven Katz, "Issues in the Separation of Judaism and Christianity after 70 C.E.: A Reconsideration," *Journal of Biblical Literature* 103 (1984), 43–76.

188. For differing interpretations, see Simon, *Verus Israel* (Paris), 234–5; Hans Joachim Schoeps, *The Jewish–Christian Argument* (New York 1963), 14–15; John Meagher, "As the Twig Was Bent: Antisemitism in Greco-Roman Earliest Christian Times," in Alan Davies, ed., *Antisemitism and the Christian Mind; The Crisis of Conscience After Auschwitz* (New York 1969), 22, 26 n43.
189. A Palestinian text from the Cairo Genizah, in Lawrence Schiffman, "At the Crossroads: Tannaitic Perspectives on the Jewish–Christian Schism," in Jeremy Cohen, ed., *Essential Papers on Judaism and Christianity in Conflict* (New York 1991), 445. See Moody, "Judaism and the Gospel of John," 84 n9; and R. Travers Herford, *Christianity in Talmud and Midrash* (New York 1975), 125–37, for other versions.
190. John 9:22, 12:42, 16:2; Schiffman, "At the Crossroads," 446; Justin and Jerome quoted by Simon, *Verus Israel* (Oxford 1986), 255.
191. Schiffman, "At the Crossroads," 448.
192. Psalm 139:21–2 is cited: "Do not I hate them, O Lord, that hate thee? And am I not grieved with those that rise up against thee? I hate them with perfect hatred: I count them mine enemies." See also Berakoth 28b–29a; TShab 13 (14), 5.
193. Num. 15:39 and *Tosefta*, Sanhedrin 13:4, 5 in Herford, *Christianity in Talmud and Midrash*, 195–9, 366–7. Schiffman argues that *Minim* were still considered Jews although their beliefs subjected them to halachic (Jewish legal) restrictions, such as the inability to serve as a cantor in the synagogue; "At the Crossroads," 444.
194. Simon, *Verus Israel*, 179–201, 254–6; Schiffman, "At the Crossroads," 446–7.
195. Kimelman, "*Birkat Ha-Minim* and the Lack of Evidence for an Anti-Christian Jewish Prayer in Late Antiquity," in Sanders, *Jewish and Christian Self-Definition*, 2:239–40. See also Katz, "Issues in the Separation of Judaism and Christianity after 70 C.E.," 43–76.
196. Shabbath 104b; Berakoth 28b–29a; Sanhedrin 43a, 67a; Gittin 56b–57a. For a summary of the Hebrew sources on Jesus, see Klausner, *Jesus of Nazareth*, 18–54.
197. *Tosefta*, Sanhedrin XIII, in Davies, *Antisemitism and the Christian Mind*, 132.
198. Amos 9:11–12 and Isa. 45:20–1, which suggest that a righteous person will help restore David's dynasty and reject idols in favor of the one "righteous God and savior." These responsibilities are also mentioned in Acts as part of the debate between Simeon (Simon Peter) and Saul (Paul) at the Council of Jerusalem (Acts 15:16–20).
199. Lev. 19:33–4; see also Exod. 22:21, Num. 15:15, 35:15, Deut. 1:16, 5:14, 10:18–19, Josh. 20:9.
200. Ruth 4:13–17. The Gospels repeat the connection, maintaining that David was an ancestor of Jesus. Matt. 1:4–6 and Luke 3:32.
201. Sanhedrin 56–60; and Maimonides, *Hilkhoth Melakhim*, 8, 11; *Sefer Hasidim*, 358. See also Schoeps, *The Jewish–Christian Argument*, 15; Jacob Katz, *Exclusiveness and Tolerance* (New York 1962), 35, 121.
202. Lev. 19:33–4.

203. As evidenced in Tertullian, Origen, Justin Martyr, Jerome, Aphraates, and others. Simon, *Verus Israel* (Eng.), 143, 173, 177, 465 n138. See also *Tractatus Adversus Judaeos*, 5:6, *PL*, 42:54.
204. Jerome, *Epistle to Titus*, 3.9 (*PL*, 26:595), quoted in Simon, *Verus Israel* (Eng.), 177, see also 230.
205. Exod. 24:12 states, "The Lord said to Moses, Come up to Me to the mountain and I will give you the stone tablets and the Torah and the Commandment which I have inscribed to instruct them." Rabbi Simeon ben Kakish, in Berakhot 5a, interprets this passage as including all the elements of the Dual Torah.
206. A brief collection of laws based on the Torah, collected about 210 C. E., and commented on in the Talmud.
207. Jacob Neusner, *Death and Birth of Judaism: The Impact of Christianity, Secularism, and the Holocaust on Jewish Faith* (New York 1987), 47.
208. Tanach Va-Yera, 5; Tanach Ki-Tissa, 34.

CHAPTER 3 MEDIEVAL VIOLENCE

1. New York 1952–69.
2. *Vallée des Pleurs* (Paris 1881).
3. Richard Southern, *Western Society and the Church in the Middle Ages* (London 1972), 16.
4. The English word, "Christendom" was first recorded in late-nineteenth-century Anglo Saxon England. King Alfred used it in 893. Judith Herrin, *The Formation of Christendom* (Princeton 1987), 8.
5. Manuel, *Broken Shaft*, 293.
6. Richard of Howden was the chronicler, in Joseph Jacobs, *The Jews of Angevin England: Documents and Records* (London 1893), 105–6; Parkes, *Church and Synagogue*, 394–400; Simonsohn, *History*, 17 n65; Robert Chazan, *Church, State, and Jew in the Middle Ages* (New York 1980), doc. 4; Simonsohn I, doc. 353.
7. Trachtenberg, *The Devil and the Jews*, 187.
8. Solomon Grayzel, "Legislation from Provincial and Local Councils Concerning the Jews," *The Church and the Jews in the XIIIth Century* (Detroit 1989), doc. 16 (hereafter referred to as Grayzel I, II).
9. Moshe Lazar, "The Lamb and the Scapegoat: The Dehumanization of the Jews in Medieval Propaganda Imagery," in Gilman and Katz, *Antisemitism in Times of Crisis*, 38.
10. Gavin Langmuir, "The Jews and the Archives of Angevin England: Reflections on Medieval Antisemitism," *Traditio* (New York 1963), 19:190–1, 232, 236.
11. See Langmuir, "Toward a Definition of Antisemitism," 207–8; David Berger, "The Jewish–Christian Debate in the High Middle Ages," in Cohen, *Essential Papers*, 486.

12. Norman Cohn, *Europe's Inner Demons* (New York 1977), 63–4, 262.
13. Trachtenberg, *The Devil and the Jews*, 161–82; Jeremy Cohen, *The Friars and the Jews* (Ithaca 1982), 47–50.
14. Langmuir, *History, Religion, and Antisemitism* (Berkeley 1990), 305; and "Toward a Definition of Antisemitism," 133.
15. Chrysostom, *Homilies against Judaizing Christians*, 1.6. 8.
16. Schreckenberg, *Die christlichen Adversus-Judaeos-Texte*, 132–7.
17. For an exhaustive summary of sources of anti-Jewish Christian material, both textual and artistic, up to the Fourth Lateran Council of 1215, see Schreckenberg, *Die christlichen Adversus-Judaeos-Texte*, and Schreckenberg, *Die christlichen Adversus-Judaeos-Texte: 11–13. Jh.* (Frankfurt 1988).
18. David Biale, *Power and Powerlessness in Jewish History* (New York 1987), 64.
19. Simonsohn., *History*, 98.
20. Simonsohn, *History*, 98 n12.
21. Jacobs, *Jews of Angevin England*, 62–3.
22. Gavin Langmuir, "*Tanquam servi*: The Change in Jewish Status in French Law About 1200," in Myriam Yardeni, ed., *Les Juifs dans l'histoire de France* (Leiden 1980), 43.
23. Alexander Broadie emphasizes the positive side of Aquinas' ambivalent opinions on the Jews. "Medieval Jewry through the Eyes of Aquinas," in G. Verbeke and D. Verhelst, eds., *Aquinas and the Problems of His Time* (The Hague 1976), 57–68.
24. Thomas Aquinas, *De Regimine principum et de regimine Judaeorum* (Turin 1924), 117; *Opuscula Omnia* (Paris 1927), 1:488. See also his *Summa Theologiae*, 11–11, Q. l0, a. 12, and 111, Q. 68, a. 10 (also called *Summa Theologica*).
25. Thomas Aquinas, "Letter to the Duchess of Brabant," in Chazan, *Church, State and Jew*, 200. The addressee may have been Louis IX's daughter, Margaret, who had married the duke of Brabant. The letter may have been written to her while the king was on Crusade, in 1270. Liebeschütz, "Judaism and Jewry in the Social Doctrine of Thomas Aquinas," n27.
26. Simonsohn I, doc. 82.
27. Gilbert Dahan, "l'article *Iudei* de la *Summa Abel* de Pierre le Chantre," *Revue des études Augustiniennes* 27 (1981), 110–11.
28. Langmuir, "Toward a Definition of Antisemitism," 139; Salo Baron, "Ghetto and Emancipation: Shall We Revise the Traditional View?" *The Menorah Journal* (June 1928), 518-19; Yosef Yerushalmi, "Response to Rosemary Ruether," in Eva Fleischner, ed., *Auschwitz: Beginning of a New Era?* (New York 1977), 98–101.
29. Joseph Reiger, "Jews in Medieval Art," in Koppel Pinson, ed., *Essays on Antisemitism* (New York 1946), 101.
30. Henry Kraus, *The Living Theatre of Medieval Art* (Philadelphia 1967), 156–8.
31. Bernhard Blumenkranz, *Le juif mediéval au Miroir de l'art chrétien* (Paris 1966), illustrations 90, 104, 117.
32. Georg Liebe, *Das Judentum in der deutschen Vergangenheit* (Leipzig 1903), illustrations 17–27.

33. Zefira Rokeah, "The State, the Church, and the Jews in Medieval England," in Almog, *Antisemitism through the Ages*, 104–11.
34. John Wasson, "The Morality Play: Ancestor of Elizabethan Drama," in Clifford Davidson et al., eds., *The Drama of the Middle Ages: Comparative and Critical Essays* (New York 1982), 320.
35. Lester Little, "The Jews in Christian Europe," in *Religious Poverty and the Profit Economy in Medieval Europe* (Ithaca 1978), reprinted in Cohen, *Essential Papers*, 277.
36. Mansi, XXIII, 22; Hefele-Leclercq V and VI; Solomon Grayzel, *The Church and the Jews in the XIIIth Century* (Philadelphia 1933); Hannah Arendt, "The Jew as Pariah: Jewish Identity and Politics in the Modern Age," (New York 1978); and Max Weber, *Wirtschaftsgeschichte* [General Economic History, orig. Munchen and Leipzig: Duncker & Humblot, 1923] (1958), 175, where his definition of the Jew is summarized as "Anyone who clung to the Jewish ritual could not be a farmer. Thus the Jews became a pariah people."
37. Marcus, *The Jew in the Medieval World*, doc. 19.
38. Dwayne Carpenter, *Alfonso X and the Jews: An Edition of and Commentary on Siete Partidas 7.24 "de los judios"* (Berkeley 1986), 5, 69, 104–5.
39. Parkes, *Church and Synagogue*, Appendix One, 385; Jones, *The Later Roman Empire*, 2:865–6.
40. See Little, "The Jews in Christian Europe," 280–1.
41. Colin Morris, *The Papal Monarchy: The Western Church from 1050 to 1250* (Oxford 1991), 45.
42. R. Straus, "The Jews in the Economic Evolution of Central Europe," *Jewish Social Studies* 3 (1941), 20; Cecil Roth, *A Short History of the Jewish People* (London 1948), 206–7.
43. Hyam Maccoby, ed., *Judaism on Trial: Jewish Christian Disputations in the Middle Ages* (East Brunswick, NJ 1982), 19–38, 89–91, 153–68, 187, 189, 197, 205; Robert Chazan, *Medieval Jewry in Northern France: A Political and Social History* (Baltimore 1973), 124–33; Chazan, *Church, State, and Jew*, 231–8; Grayzel, *The Church and the Jews*, 29–33; Joel Rembaum, "The Talmud and the Popes: Reflections on the Talmud Trials of the 1240s," *Viator: Medieval and Renaissance Studies* 13 (1982), 203–23; Werner Keller, *Diaspora: The Post-Biblical History of the Jews* (New York 1969), 284–8, 295, 297; Marcus, *The Jew in the Medieval World*, 170–3; Cecil Roth, "The Disputation of Barcelona," in Roth, ed., *Gleanings* (New York 1967), 34–61.
44. Grayzel, *The Church and the Jews*, 340.
45. Grayzel I, 277–9 n3; Simonsohn I, doc. 309.
46. Cohen, *The Friars and the Jews*, 174–95.
47. Marcus, *The Jew in the Medieval World*, 149.
48. John of Joinville, *The Life of Louis* (New York 1955), 35–6.
49. Chazan, *Medieval Jewry in Northern France*, 100–2.
50. Chazan, *Medieval Jewry in Northern France*, 102–3.
51. Keller, *Diaspora*, 225.
52. Joel Rembaum, "Medieval Christianity Confronts Talmudic Judaism," 376.

53. Trachtenberg, *The Devil and the Jews*, chapter 15. Exod. 22:18 required that "You shall not permit a sorceress to live."
54. Morris, *The Papal Monarchy*, 350–3.
55. H.R. Trevor-Roper, *The European Witch-Craze of the 16th and 17th Centuries* (New York 1967), 139.
56. Isaiah Shachar, *The Judensau* (London 1974), 48.
57. Trevor-Roper, *The European Witch-Craze*, 112–13; Shachar, *The Judensau*, 48.
58. Pierre de l'Ancre, *L'incrédulité et mescréances du sortilège pleinement convaincues* (Paris 1622), 8:446–8.
59. Marcus, *The Jew in the Medieval World*, 50.
60. Buxtorf. Grayzel, *The Church and the Jews*, 301, n3.
61. Grayzel, *The Church and the Jews*, 333.
62. Grayzel, *The Church and the Jews*, 109.
63. See Grayzel, *The Church and the Jews*, 72–5, for the complete list.
64. Trachtenberg, *The Devil and the Jews*, 187.
65. R. Edelmann, "Ahasuerus, the Wandering Jew: Origin and Background," in Galit Hasan-Rokem and Alan Dundes, eds., *The Wandering Jew: Essays in the Interpretation of a Christian Legend* (Bloomington, IN 1986), 7.
66. E. Isaac-Edersheim, "Ahasver: A Mythic Image of the Jew," in Hasan-Rokem and Dundes, *The Wandering Jew*, 205.
67. Blumenkranz, *Juifs et Chrétiens*, 382–3; Marcel Bulard, *Le Scorpion* (Paris 1935); Allan and Helen Cutler, *The Jew as Ally of the Muslim* (Notre Dame, IN 1986).
68. Matt. 24:3–31; Mark 13:3–37. See also 1 John 2:18, 2:22, 4:3, and 2 John 7.
69. Norman Cohn, *Pursuit of the Millennium* (New York 1980), 77–8, 340 nn77–8.
70. Lazar, "The Lamb and the Scapegoat," 47, 55, and illustration 70.
71. Bousset, *The Antichrist Legend*, 158–74; Hyam Maccoby, *The Sacred Executioner: Human Sacrifice and the Legacy of Guilt* (New York 1982), 172. See also M. Reeves, *The Influence of Prophecy in the Later Middle Ages* (Oxford 1969); and B. McGinn, *Visions of the End: Apocalyptic Traditions in the Middle Ages* (New York 1979).
72. John Wright, *The Play of the Antichrist* (Toronto 1967), 24–40.
73. *Contra Perfidiam Judaeorum*, in Jacobs, *Jews of Angevin England*, 182; and Jeremy Cohen, "Traditional Prejudice and Religious Reform," in Gilman and Katz, *Antisemitism in Times of Crisis*, 90.
74. Peter the Venerable, "Tractatus Adversus Judeorum Inveteratam Duritiem," *PL*, 189:507, 648–50.
75. *Ordinary of Vincent of Chalons*, in Roth, *Gleanings*, 81–6.
76. *The Chronicle of Gaufredi Vosiensis*, in Martin Bouquet et al., eds., *Recueil des historiens des Gaules et de la France* (Paris 1737–1904), 12:194; Henry Milman, *The History of the Jews in England* (London 1830, 1909), 2:302.
77. Charles Lehrmann, *The Jewish Element in French Literature* (Cranbury, NJ 1961), 57 n3; Bernhard Blumenkranz, *Juifs et Chrétiens dans le monde occidental* (Paris 1960), 382–3; Keller, *Diaspora*, 168; Gustave Saige, *Les juifs de Languedoc antièrement le XIVe siècle* (Paris 1881), 64, in Kraus, *The Living Theatre of Medieval Art*, 148; Little, "The Jews in Christian Europe," 282.

78. Roth, *Gleanings*, 6–15, 81 n1.
79. Quoted in Kertzer, *Popes against the Jews*, 74.
80. Edward Synan, *The Popes and the Jews in the Middle Ages* (New York 1965), 146; Cecil Roth, *A History of the Jews of Italy* (Philadelphia 1946), 386–7; Sam Waagenaar, *The Pope's Jews* (La Salle, IL 1974), 137–43; Nicolas Bénard, *Le Voyage de Hierusalem et autres lieux* . . . (Paris 1621), 520; and Myriam Yardeni, *Anti-Jewish Mentalities in Early Modern Europe* (Lanham, MD 1990), 154 n57.
81. See Pelikan, *The Spirit of Eastern Christendom*, 201; Carol Krinsky, *Synagogues of Europe: Architecture, History, Meaning* (Cambridge, MA 1985), 38; Jaroslav Pelikan, *The Christian Tradition: The Emergence of the Catholic Tradition, 100–600* (Chicago 1971), 23, 66–7; David Efroymson, "The Patristic Connection," in Alan Davies, ed., *Antisemitism and the Foundations of Christianity* (New York 1979), 114.
82. Edward Peters, *Inquisition* (Berkeley 1989), 22.
83. Efroymson, *Tertullian's Anti-Judaism*, 109.
84. Ambrose, "In Psalmum XXXIX Enarratio," *PL*, 14:1062.
85. Revel-Neher, *The Image of the Jew in Byzantine Art*, 30.
86. J.-B. Palanque, *Saint Ambroise et l'Empire romain* (Paris 1933), 366; Williamson, *Has God Rejected His People?* 89; Effroymson, *Tertullian's Anti-Judaism and Its Role in His Theology*; Wilken, *Judaism and the Early Christian Mind*, 46.
87. *Codex Theodosianus*, 16:5:44.
88. Blumenkranz, *Le juif medieval au miroir de l'art chretien*, 77; Emil Friedberg, ed., *Corpus Juris Canonici* (Leipzig 1881), 1:997–8.
89. Quoted by Peters, *Inquisition*, 42.
90. See Cohen, *The Friars and the Jews*; Walter Ullman, *A History of Political Thought* (Baltimore 1970), l05; Cohn, *Pursuit of the Millenium*, 101 and *Warrant for Genocide* (Chico, CA 1981), 15–25; Adolf Leschnitzer, *The Magic Background of Modern Antisemitism* (New York 1956), 98; Trachtenberg, *The Devil and the Jews*, 175, 249.
91. Trachtenberg, *The Devil and the Jews*, 180–1.
92. Trachtenberg, *The Devil and the Jews*, 180–1.
93. Blumenkranz, *Le juif mediéval au miroir de l'art chrétien*, 77; Trachtenberg, *The Devil and the Jews*, 174.
94. Cohen, *The Friars and the Jews*, 14, 41, 45, 48, n37.
95. Cohen, *The Friars and the Jews*, 14, 21, 124–5, 145; Cohn, *Warrant for Genocide*, 15–25, 307–18 and *Pursuit of the Millenium*, 99–101; Marcus, *The Jew in the Medieval World*, 145–50.
96. Cohen, "Traditional Prejudice and Religious Reform," 97.
97. David Christie-Murray, *A History of Heresy* (Oxford 1989), 108; Richardson, *The Jews of Angevin England*, 35.
98. Aquinas, *Summa Theologiae*, 2a2ae.10, 10–12; 2a2ae.14, 3; 3a, 47, 5–6; 1a2ae, 102, 6.8.
99. Aquinas, *Summa Theologiae*, 2a 2ae, 14, 3.

100. Cecil Roth, "The Medieval Conception of the Jew," in Israel Davidson, ed., *Essays and Studies in Memory of Linda R. Miller* (New York 1938), 171–90; Jeremy Cohen, "The Jews as the Killers of Christ in the Latin Tradition, from Augustine to the Friars," *Traditio* 39 (1983), 1.
101. Simonsohn, *History*, 313–15; Cohen, *The Friars and the Jews*.
102. Trachtenberg, *The Devil and the Jews*, 175, 227 n18; Cohen, *The Friars and the Jews*, 231–6. See Berthold von Regensburg, *Berthold von Regensburgs Deutsche Predigten* (Jena 1924), Sermon 42.
103. Hans Liebeschütz, "The Crusading Movement in Its Bearing on the Christian Attitude towards Jewry," *Journal of Jewish Studies* 10 (1959), 97–111.
104. Morris, *The Papal Monarchy*, 152; Hans Mayer, *The Crusades* (New York 1988), 14.
105. Morris, *The Papal Monarchy*, 339. In the previous century, Jews of the Byzantine Empire, including Byzantine-controlled Italy, had been forced to convert following rumors about damage to the Holy Sepulchre. Simonsohn, *History*, 12.
106. *Chronicle* 3:7, *PL*, Vol. 142, quoted by Malcolm Hay, *Thy Brothers' Blood*, 37.
107. Abraham Habermann, *Sefer Gezerot*, in Chazan, *Church, State, and Jew*, 293–4; Little, "Jews in Christian Europe," 281.
108. Simonsohn I, doc. 37, see also docs. 36 and 38.
109. Simonsohn I, doc. 35, note.
110. Morris, *The Papal Monarchy*, 150–2.
111. Norman Golb, "New Light on the Persection of French Jews at the Time of the First Crusade," in Robert Chazan, ed., *Medieval Jewish Life* (New York 1976), 325–6.
112. Golb, "New Light on the Persection of French Jews," 322–3; Cohn, *Pursuit of the Millennium*, 68.
113. Cohn, *The Pursuit of the Millennium*, 61.
114. Solomon bar Simson, quoted in Robert Chazan, *European Jewry and the First Crusade* (Berkeley 1987), 68.
115. Simonsohn I, doc. 42. See also Pakter, *De His Qui Foris Sunt*, 290; Parkes, *The Jew in the Medieval Community*, 79–80, 85.
116. Simonsohn, *History*, 277–8.
117. Simonsohn I, doc. 44.
118. Mayer, *The Crusades*, 8.
119. Morris, *The Papal Monarchy*, 149.
120. Morris, *The Papal Monarchy*, 150–2.
121. Mayer, *The Crusades*, 9.
122. Golb, "New Light on the Persection of French Jews," 318–29; Ivan Marcus, "From Politics to Martyrdom," *Prooftexts* 2 (1982), 40–52; and Robert Chazan, "The Facticity of Medieval Narrative," *AJS Review* 16 (1991), 31–56.
123. Mayer, *The Crusades*, 8, 23, 30–3, 36; Morris, *The Papal Monarchy*, 150–2; Cohn, *Pursuit of the Millennium*, 61.
124. Mainz Anonymous, in Chazan, *European Jewry*, 226; and Shlomo Eidelberg, ed., *The Jews and the Crusaders: The Hebrew Chronicles of the First and Second Crusades* (Madison 1977), 100.

125. Morris, *The Papal Monarchy*, 355; Mayer, *The Crusades*, 40.
126. Cohn, *Pursuit of the Millennium*, 62–3.
127. Mayer, *The Crusades*, 55.
128. Golb, "New Light on the Persection of French Jews."
129. Riley-Smith, *The First Crusade and the Idea of Crusading*, 50–1; Steven Runciman, *History of the Crusades: The First Crusade* (London 1953), 1:137. Régine Pernoud argues that the People's Crusade was most likely in Hungary or the Balkans when the murders of Jews took place. *The Crusades* (Edinburgh 1963), 100–01.
130. Riley-Smith, *The First Crusade and the Idea of Crusading*, 52. The sermon at Jerusalem from the *Historia* of Baldric, bishop of Bourgueil, written in 1108, quoted in Riley-Smith, *The First Crusade and the Idea of Crusading*, 48–9.
131. Riley-Smith, *The First Crusade and the Idea of Crusading*, 49.
132. Riley-Smith, *The First Crusade and the Idea of Crusading*, 50. Golb believes that it was the French contingents who massacred Jews at Regensburg and Prague. Golb, "New Light on the Persection of French Jews," 315.
133. Whereas other scholars such as Riley-Smith regard these Crusaders as roving bands, Chazan sees them as full-fledged armies. Chazan, *European Jewry*, 63–4.
134. Baldric of Dol, in Golb, "New Light on the Persection of French Jews," 39–40 n78.
135. Mayer, *The Crusades*, 39–43, 57.
136. Eidelberg, "General Introduction," *The Jews and the Crusaders*, 5–6.
137. Robert Chazan, "The Hebrew First-Crusade Chronicles," *Revue des études juives: Historia Judaica* 33 (January–June 1974), 248.
138. Guibert of Nogent, *De Vita Sua*, II, 5, in Bouquet et al., *Recueil des historiens des Gaules et de la France*, 12:240; and *The Chronicle of Solomon bar Simson*, in Eidelberg, *The Jews and the Crusaders*, 22; see also Georges Bourgin, ed., *Guibert de Nogent, Histoire de sa vie* (Paris 1907); John Benton, ed., *Self and Society in Medieval France: The Memoirs of Guibert of Nogent* (New York 1970).
139. Mainz Anonymous, in Chazan, *European Jewry*, 229; Golb, "New Light on the Persection of French Jews," 307–10.
140. Simonsohn I, doc. 82.
141. Ben-Zion Dinur, ed., *Yisrael ba-Golah* (Tel Aviv 1958–72), cited by Chazan, *European Jewry*, 198.
142. For Eugenius III's bull, *Quantum praedecessores nostri*, see Chazan, *European Jewry*, 179–80; Grayzel I, docs. 1, 25, 26, 28, 58, 80, 90, 91, and 313.
143. Albert of Aix, *History of the Crusades*, Book I, in M. Guizot, *Collection des Mémoires Relatifs à l'Histoire de France* (Paris 1824; New York 1969), 20:44. See also Riley-Smith, *The First Crusade and the Idea of Crusading*, 52–3.
144. Mayer, *The Crusades*, 41.
145. Riley-Smith, *The First Crusade and the Idea of Crusading*, 52.
146. Riley-Smith, "The First Crusade and the Persecution of the Jews," in W.J. Sheils, ed., *Persecution and Toleration* (Papers of the Ecclesiastical History Society 1984), 51–6; Chazan, *European Jewry*, 57.
147. Riley-Smith, "The First Crusade and the Persecution of the Jews," 57; Chazan, *European Jewry*, 81.

148. Henri Peyre, "The Influence of Ideas on the French Revolution," *The Journal of the History of Ideas* 10 (1949), 72.
149. Morris, *The Papal Monarchy*, 152.
150. Chazan, *European Jewry*, 204.
151. Riley-Smith, "The First Crusade and the Persecution of the Jews," 51–6; Chazan, *European Jewry*, 80–1.
152. See Langmuir, *History, Religion, and Antisemitism*, 291 and "Toward a Definition of Antisemitism," 97. See also, Langmuir, *History, Religion, and Antisemitism*, chapters 7 and 9.
153. Albert of Aix, *History of the Crusades*, Book I, 20:38. See also Albert of Aix, *Liber Christianae expeditionis in Recueil des historiens des croisades, historiens occidentaux* 4:292, in Chazan, *European Jewry*, 64.
154. Matt. 16:24. Ironically, this statement has also been used to justify decent behavior toward Jews.
155. An anonymous chronicler of Mainz in Adolf Neubauer and Moritz Stern, eds., *Hebräische Berichte über die Judenverfolgungen während der Kreuzzüge* (Berlin 1892), 176, 169; Chazan, "The Hebrew First-Crusade Chronicles," 249–50, 253; Bernold, *Chronicle in Monumenta Germaniae Historica, Scriptores* (Hannover 1826–96), 5:465.
156. See Ekkehard of Aura, *Hierosolymita*, cap. xii.
157. Langmuir, "Toward a Definition of Antisemitism," 97.
158. See August Krey, ed., *The First Crusade: The Accounts of Eye-Witnesses and Participants* (Princeton 1921), 53.
159. Solomon bar Simson, in Chazan, *European Jewry*, 245.
160. Mainz Anonymous, in Chazan, *European Jewry*, 239.
161. Golb, "New Light on the Persecution of French Jews," 19.
162. The First Crusaders consisted of Rhinelanders, Swabians, French, English, Flemish, Lorrainers, Saxons, Bohemians, Bavarians. Riley-Smith, "The First Crusade and the Persecution of the Jews," 51–2.
163. Golb, "New Light on the Persection of French Jews," 322.
164. On unpremeditated violence, see Chazan, *European Jewry*, 56–7.
165. Riley-Smith, "The First Crusade and the Persecution of the Jews," 68.
166. Solomon bar Simson, in Chazan, *European Jewry*, 68.
167. Mainz Anonymous, in Chazan, *European Jewry*, 226–7.
168. Chronicle of Solomon bar Simson, in Eidelberg, *The Jews and the Crusaders*, 28.
169. Quoted by Chazan, *European Jewry*, 59–61.
170. Eidelberg, *The Jews and the Crusaders*, 102; Chazan, *European Jewry*, 57.
171. Albert of Aix, in Krey, *The Crusades*, 54.
172. Chazan, *European Jewry*, 70.
173. Golb, "New Light on the Persection of French Jews," 24.
174. Riley-Smith, *The First Crusade and the Idea of Crusading*, 53; Riley-Smith, *The Crusades* (New Haven 1987), 17. This kind of behavior characterized the Nazis as well. See chapter 11.
175. See Mainz Anonymous and Solomon bar Simson in Chazan, *European Jewry*, 240, 245, 260, 274, 288. See also "The Chronicle of Rabbi Eliezer bar Nathan," in Eidelberg, *The Jews and the Crusaders*, 85.

176. Golb, "New Light on the Persecution of French Jews," 319–20 n58. See also Chazan, *European Jewry*, 289.
177. Reported by Baldric of Bourgeuil, see Riley-Smith, *The First Crusade and the Idea of Crusading*, 55.
178. Mayer, *The Crusades*, 50; Cohn, *Pursuit of the Millennium*, 66–7.
179. Riley-Smith, "The First Crusade and the Persecution of the Jews," 51–6; Chazan, *European Jewry*, 63.
180. *Guibert de Nogent: Histoire de sa vie* (Paris 1907), 118; Chazan, *European Jewry*, 54; Riley-Smith, *The First Crusade and the Idea of Crusading*, 50–2; Golb, "New Light on the Persection of French Jews," 319–20, 329, 333.
181. Mainz Anonymous, translated by Chazan, *European Jewry*, 225.
182. Solomon bar Simson, translated by Chazan, *European Jewry*, 244.
183. Richard of Poitiers, quoted by Riley-Smith, *The First Crusade and the Idea of Crusading*, 54.
184. See Simson, *Chronicle of Solomon bar Simson* and Mainz Anonymous, *Narrative of the Old Persecutions* in Eidelberg, *The Jews and the Crusaders*, 125–6. See also Mainz Anonymous, in Chazan, *European Jewry*, 226.
185. Chazan, *European Jewry*, 247.
186. Norman Golb, "New Light on the Persection of French Jews," 326–7; Chazan, *European Jewry*, 53.
187. John Dalberg-Acton, "Inaugural Lecture on the Study of History," in *Lectures on Modern History* (London 1921), 24–5.
188. Raymond of Aguilers, "Historia Francorum qui ceperunt Jerusalem," in *Recueil des historiens des Croisades: Historiens occidentaux* (Paris 1844–95), 3:300, in Cohn, *Pursuit of the Millennium*, 68.
189. Riley-Smith, *The First Crusade and the Idea of Crusading*, 53–4; Chazan, *European Jewry*, 269, 289–90.
190. Mainz Anonymous and Solomon bar Simson, in Chazan, *European Jewry*, 227, 245.
191. Solomon bar Simson, in Chazan, *European Jewry*, 246–7, 251, 253, 269. Some bishops may have been bribed by the Jews.
192. Mainz Anonymous in Chazan, *European Jewry*, 227–8. See also Chazan, "The Hebrew First-Crusade Chronicles," 249.
193. Solomon bar Simson, in Chazan, *European Jewry*, 253, 260–1.
194. Mainz Anonymous and Solomon bar Simson, in Chazan, *European Jewry*, 233, 249.
195. Solomon bar Simson, Chazan, *European Jewry*, 274.
196. Solomon bar Simson, in *Narrative of the Old Persecutions*, in Eidelberger, *The Jews and the Crusaders*, 101–2. See also Chazan, *European Jewry*, 288.
197. Rabbi Ephraim bar Jacob of Bonn, in Neubauer and Stern, *Hebräische Berichte*, 187–8. Otto of Freising quoted in Chazan, *European Jewry*, 170.
198. Cohn, *Pursuit of the Millennium*, 69–70.
199. See Graetz, *The History of the Jews*, 3:353–6.
200. Guibert de Nogent and Richard of Poitiers, in Cohn, *Pursuit of the Millennium*, 70.

201. Rabbi Ephraim bar Jacob of Bonn, in Neubauer and Stern, *Hebräische Berichte*, 187–8. Otto of Freising quoted in Chazan, *European Jewry*, 170.
202. Eidelberg, *The Jews and the Crusaders*, 127–8.
203. Graetz, *The History of the Jews*, 3:351–2; Vamberto Morais, *A Short History of Antisemitism* (New York 1976), 104.
204. Chazan, *European Jewry*, 177–8.
205. Chazan, *Church, State, and Jew*, 101–4.
206. Rabbi Ephraim bar Jacob of Bonn, in Neubauer and Stern, *Hebräische Berichte*, 188. See also Milman, *History of the Jews*, 2:310.
207. Chazan, *Church, State, and Jew*, 104–5.
208. Bernard of Clairvaux, "Epistola CCCLXIII (946)," *PL*, 182:567. In Paul and Augustine, the conversion of all the Jews was a prerequisite to the Second Coming. See Rom. 11:25–6, and Augustine, *The City of God*, 20:30.
209. "Bernard's Letter to the People of England," in Chazan, *Church, State, and Jew*, 101–4.
210. Bernard of Clairvaux, *Bernard's Sermons for the Seasons and the Principal Festivals of the Year* (Westminster, MD 1950), 1:379.
211. *De Consideratione* I, 4, quoted by Jean-Pierre Torrell, "Les Juifs dans l'oeuvre de Pierre le Vénérable," *Cahiers de civilisation médiévale* 30 (1987), 342 n58.
212. Bernard, *Sermones super Cantica Canticorum*, 60.4, in David Berger, "The Attitude of St. Bernard of Clairvaux toward the Jews," *American Academy for Jewish Research, Proceedings* (New York 1973), 96.
213. Bernard, *Bernard's Sermons*, 2:149, in Berger, "The Attitudes of St. Bernard of Clairvaux toward the Jews," 104.
214. Bouquet et al., *Recueil des historiens des Gaules et de la France*, 15:606; Chazan, *Church, State, and Jew*, 103.
215. Berger, "The Attitude of St. Bernard of Clairvaux toward the Jews," 101–2.
216. Stroll, *The Jewish Pope*, 166.
217. Stroll, *The Jewish Pope*, 159–62, 177, 181.
218. Chazan, *Medieval Jewry in Northern France*, 46; see also Amos Funkenstein, "Changes in the Patterns of Christian Anti-Jewish Polemics in the Twelfth Century," *Zion* 33 (1968), 125–44; Torrell, "Les Juifs dans l'oeuvre de Pierre le Vénérable," 336.
219. Langmuir, "Toward a Definition of Antisemitism," 207–8; Berger, "The Jewish–Christian Debate," 486.
220. *PL*, 189:661.
221. Peter the Venerable, *The Letters of Peter the Venerable*, ed. Giles Constable (Cambridge, MA 1967), 2:185. See Langmuir's citations in "Toward a Definition of Antisemitism," 202–3, 383.
222. Synan, *The Popes and the Jews*, 76.
223. Constable, *The Letters of Peter the Venerable*, Letter 130.
224. Guibert de Nogent, *De Vita sua*, 2:5 (Paris 1981), 246.
225. Reported by Guibert de Nogent, see Torrell, "Les juifs dans l'oeuvre de Pierre le Vénérable," 339.
226. Constable, *The Letters of Peter the Venerable*, Letter 130.

227. See *Liber contra sectam siue haeresim Sarracenorum* in James Kritzeck, ed., *Peter the Venerable and Islam* (Princeton 1964), 265; see also Torrell, "Les juifs dans l'oeuvre de Pierre le Vénérable," 339 n45.
228. Peter the Venerable, "Epistola XXXVI," *PL*, 189:367–8; Constable, *The Letters of Peter the Venerable*, Letter 130.
229. Psalm 139, line 19.
230. Peter the Venerable, "Epistola XXXVI," *PL*, 189:367–8; Constable, *The Letters of Peter the Venerable*, Letter 130.
231. Georges Duby, "Le budget de l'abbaye de Cluny entre 1080 et 1144," *Annales* 7 (1952), 155–72, in Kraus, *The Living Theatre of Medieval Art*, 147.
232. Constable, *The Letters of Peter the Venerable*, Letter 130. That there was a widespread revulsion at the idea of Jewish ownership of Church property can be seen in English law. See King Richard's and King John's charters of 1190 and 1201. Chazan, *Church, State, and Jew*, docs. 5 and 6, 68, 78.
233. Peter the Venerable, "Epistola XXXVI," *PL*, 189:367–8; Constable, *The Letters of Peter the Venerable*, Letter 130.
234. Edouard Drumont, *La France juive* (Paris 1885), 1:526.
235. Chazan doubts the accuracy of Ephraim of Bonn's report. Chazan, *European Jewry*, 182–3, 187–8.
236. Peter the Venerable, "Tractatus Adversus Judaeorum Inveteram Duritiem," *PL*, 189:507–650.
237. Peter the Venerable, "Tractatus Adversus Judaeorum Inveteram Duritiem," *PL*, 189:560.
238. Peter the Venerable, *Contra Petrobrusianos hereticos*, cited by Torrell, "Les Juifs dans l'oeuvre de Pierre Vénérable," 338.
239. Peter the Venerable, "Tractatus Adversus Judaeorum Inveteratam Duritiem," *PL*, 189:602.
240. Peter the Venerable, "Tractatus Adversus Judaeorum Inveteratam Duritiem," *PL*, 189:550–1, 602, 649–50.
241. Chazan, *Medieval Jewry in Northern France*, 45–6.
242. Chazan, *Medieval Jewry in Northern France*, 47.
243. *Rabbenu* means "our teacher" and is a term of greater respect than *rabbi*. Tam's brother, Rashbam (Rabbi Samuel ben Meir), was another great Jewish Talmudist of the period.
244. Rabbi Shlomo ben Isaac of Troyes, or Rashi (d. 1105), was perhaps the most renowned Jewish scholar and religious authority of the Middle Ages. He encouraged Jews who had been forcibly converted (*anusim*, or reluctant apostates) to return to Judaism. Roth, *A History of the Marranos*, 4.
245. Ephraim ben Jacob, *A Book of Historical Records*, in Marcus, *The Jew in the Medieval World*, doc. 61.
246. As recorded by Ephraim of Bonn, quoted in Yosef Yerushalmi, *Zakhor*, 49.
247. Mayer, *The Crusades*, 135.
248. Simonsohn, *History*, 48.
249. "The Report of Rabbi Elazar ben Judah," in Chazan, *Church, State, and Jew*, doc. 8, 117–22. See also Robert Chazan, "Emperor Frederick I, the Third Crusade, and the Jews," *Viator* 8 (1970), 83–93.

250. Chazan, *European Jewry*, 172–3.
251. R.B. Dobson, *The Jews of Medieval York and the Massacre of 1190* (York 1974), 19; Simonsohn I, doc. 63.
252. According to the chronicles of Saint-Denis, in Béatrice Philippe, *Etre Juif dans la société française* (Paris 1979), 18.
253. Bouquet et al., *Recueil des historiens des Gaules et de la France*, 18:263.
254. Habermann, *Sefer Gezerot*, 162–3, 304–6.
255. Habermann, *Sefer Gezerot*, 163–5.
256. Dobson, *The Jews of Medieval York*, 20.
257. William of Norwich in 1144, Harold of Gloucester in 1168, Robert of Bury St Edmunds in 1181, Adam of Bristol in 1183.
258. Dobson, *The Jews of Medieval York*, 23.
259. "The History of William of Newburgh," in Chazan, *Church State, and Jew*, 158–61.
260. See John Appleby, *The Chronicle of Richard of Devizes* (London 1963).
261. Robert Levine, "Why Praise Jews: Satire and History in the Middle Ages," *Journal of Medieval History* 12 (1986) 292–4.
262. Milman, *History of the Jews*, 2:347.
263. Dobson, *The Jews of Medieval York*, 24–5.
264. Jacobs, *Jews of Angevin England*, 133–4.
265. Dobson, *The Jews of Medieval York*, 18, 26.
266. Dobson, *The Jews of Medieval York*, 21, 26.
267. Dobson, *The Jews of Medieval York*, 28.
268. Chazan, *European Jewry*, 190. For William of Newburgh's chronicle, see Richard Howlett, ed., *Chronicles of the Reign of Stephen, Henry II, and Richard I* (London 1884–89), 1:294–312.
269. Dobson, *The Jews of Medieval York*, 32.
270. William of Newburgh in Jacobs, *Jews of Angevin England*, 115–16.
271. William of Newburgh in Jacobs, *Jews of Angevin England*, 122.
272. Dobson, *The Jews of Medieval York*, 29.
273. Dobson, *The Jews of Medieval York*, 33, 36–7.
274. Dobson, *The Jews of Medieval York*, 26.
275. Morris, *The Papal Monarchy*, 478, 484.
276. Revel-Neher, *The Image of the Jew in Byzantine Art*, 17.
277. Graetz, *The History of the Jews*, 3:502–3; Morris, *The Papal Monarchy*, 446.
278. Golb, "New Light on the Persection of French Jews," 295.
279. Gui Alexis Lobineay, *Histoire de Bretagne* (Paris 1707), 1:235.
280. Gautier de Coincy, *Les Miracles de la Sainte Vierge*, in Poliakov, *The History of Antisemitism*, 1:53–4.
281. Lehrmann, *The Jewish Element in French Literature*, 47–8.
282. Grayzel I, doc. 70.
283. Simonsohn I, doc. 154; Grayzel I, doc. 87.
284. Simonsohn I, doc. 154; Grayzel I, doc. 87.
285. Simonsohn, *History*, 51.
286. Simonsohn I, doc. 155; Grayzel I, doc. 88.
287. Grayzel I, 229 n7, 226 n6.

288. J.M. Vidal, "L'Emeute des Pastoureaux en 1320," in *Annales deLouis des Français* 3 (1898), 138.
289. Bernard Gui, *Vita Joannis XXII*, in E. Baluze, *Vitae paparum Avinoniensium* (Paris 1814–1937), 1:161–3; John, Canon ofVictor, in Baluze, *Vitae paparum*, 1:128–30; Solomon Ibn Verga, *Shebet Yehuda* (Hanover 1856), 4–6 in Cohn, *The Pursuit of the Millennium*, 103–4.
290. Yitzhak Baer, *A History of the Jews in Christian Spain* (Philadelphia 1978), 2:15; Chazan, *Church, State and Jew*, 181–3.
291. Graetz, *The History of the Jews*, 4:57–8.
292. Simonsohn, *History*, 453.
293. Chazan, *European Jewry*, 220–1.
294. Lev. 18:5. The Jews were required to choose life over death. An act called *pikku'ah nefesh*, or regard for human life. See Sanhedrin 74 and David Roskies, *Against the Apocalypse* (Cambridge, MA 1984), chapter 2.
295. Lev. 22:32.
296. Before ten or more people.
297. As embodied in idolatry, unlawful sexual relations, or murder.
298. Lawrewnce Langer's phrase applied by him to the lack of real moral choice forced on the Jews of the Holocaust.
299. Writing in 1140, the Jewish chronicler Solomon bar Simson and the anonymous Mainz chronicler, in Neubauer and Stern, *Hebräische Berichte*, 87–9, 170, 6–8.
300. Chazan, *European Jewry*, 255. See also Chronicle of Solomon bar Simson, in Eidelberg, *The Jews and the Crusaders*, 32.
301. An anonymous chronicler of Mainz in Neubauer and Stern, *Hebräische Berichte*, 176, 169; Chazan, "The Hebrew First-Crusade Chronicles," 249–50, 253.
302. Bernold, *Chronicle in Monumenta Germaniae Historica, Scriptores*, 5:465.
303. Solomon bar Simson, in Chazan, *European Jewry*, 257.
304. Mainz Anonymous, in Chazan, *European Jewry*, 231.
305. Mainz Anonymous and Solomon bar Simson, in Chazan, *European Jewry*, 235, 240, 252–3.
306. Katz, *Exclusiveness and Tolerance*, 88–92. This phenomenon was also new to the eleventh century.
307. Mainz Anonymous, in Chazan, *European Jewry*, 237.
308. Mainz Anonymous and Solomon bar Simson, in Chazan, *European Jewry*, 238, 255, 259.
309. Jewish chroniclers, Mainz Anonymous and Solomon bar Simson, in Chazan, *European Jewry*, 66–9, 225, 230–1, 243.
310. Solomon bar Simson, in Chazan, *European Jewry*, 243.
311. Anna Abulafia, "Invectives against Christianity in the Hebrew Chronciles of the First Crusade," in Peter Edbury, ed., *Crusade and Settlement* (Cardiff, Eng. 1985), 66–7.
312. Solomon bar Simson, in Chazan, *European Jewry*, 262; and Eidelberg, *The Jews and the Crusaders*, 32, 38.

313. Much of this material may have originated in the *Toledot Jeshu*. Invective like Christianity was "error," however, came from an older talmudic tradition. Katz, *Exclusiveness and Tolerance*, chapter 3.
314. Marcus, "From Politics to Martyrdom," 469–83.
315. *Chronicle of Solomon bar Simson*, in Eidelberger, *The Jews and the Crusaders*, 25–6.
316. *Chronicle of Solomon bar Simson*, in Eidelberg, *The Jews and the Crusaders*, 33. See also Roskies, *Against the Apocalypse*, 44.
317. Based on Shlomo Eidelberg's translation of *The Chronicle of Solomon bar Simson*; see also Solomon bar Simson, in Chazan, *European Jewry*, 252, 256.
318. See Julius Streicher, "Special Ritual-Murder Edition," *Der Stuermer* (May 1934); Randall Bytwerk, *Julius Streicher* (New York 1983), 127–30, 199–200.
319. Langmuir, "Toward a Definition of Antisemitism," 127.
320. Maccoby, *The Sacred Executioner*, 147; Roth, *A Short History of the Jewish People*, 102; Trachtenberg, *The Devil and the Jews*, 129, 148.
321. Maccoby, *The Sacred Executioner*, 155; see Cecil Roth, "The Feast of Purim and the Origins of the Blood Accusation," *Speculum* 8 (1933), 525; Chazan, *Medieval Jewry in Northern France*, 49.
322. Magdalene Schultz, "The Blood Libel: A Motif in the History of Childhood," in Alan Dundes, ed., *The Blood Libel Legend* (Madison 1991), 273–303.
323. Morris, *The Papal Monarchy*, 356; Shachar, *The Judensau*, 58–60.
324. Roth, "The Feast of Purim," 520–6.
325. Ernest Rappaport, "The Ritual Murder Accusation: The Persistence of Doubt and the Repetition Compulsion," in Dundes, *The Bood Libel Legend*, 326.
326. R. Po-chia Hsia, "Jews as Magicians in Reformation Germany," in Gilman and Katz, *Antisemitism in Times of Crisis*, 119.
327. Frederick II's Charter of 1236, King James II of Aragon's Letter of 1294, in Chazan, *Church, State, and Jew*, 114–16, 123–68.
328. Hsia, "Jews as Magicians in Reformation Germany," 118–19.
329. See Streicher, "Special Ritual-Murder Edition"; Bytwerk, *Julius Streicher*, 127–30, 199–200.
330. Trachtenberg, *The Devil and the Jews*, 148.
331. Langmuir, "Toward a Definition of Antisemitism," chapter 9.
332. A second case occurred in 1168, Harold of Gloucester.
333. Langmuir, "Toward a Definition of Antisemitism," 307.
334. See Rokeah, "The State, the Jews, and the Church in Medieval England," 104–11.
335. *The Life and Miracles of St. William of Norwich by Thomas of Monmouth*, quoted in Chazan, *Church, State and Jew*, 142–5.
336. Langmuir, "Toward a Definition of Antisemitism," 307.
337. Richard of Pointoise. Morris, *The Papal Monarchy*, 356.
338. Chazan, *Medieval Jewry in Northern France*, 37; Morris, *The Papal Monarchy*, 356; Chazan, *Church, State, and the Jew*, 114–17; Philippe, *Etre Juif dans la société française*, 18.
339. Philipe, *Etre Juif dans la société française*, 18.

340. Marcus, *The Jew in the Medieval World*, doc. 5; Chazan, *Medieval Jewry in Northern France*, 44.For the Jewish reports, see Chazan, *Church, State, and the Jew*, doc. 8, 114–17.
341. Gen. 22; Deut. 18:10; Jer. 2:35–35, 19:4–6.
342. Lev. 3:17, 7:26, 17:10–4; Deut. 12:15–16, 23–5.
343. Ephrem the Syrian, *Hymns of Unleavened Bread* 19:11, 12, 16.
344. Trachtenberg, *The Devil and the Jews*, 6, 47–50, 116, 149–50, 227–8; Levi, "Le Juif de la legende," 249–51; Poliakov, *The History of Antisemitism*, 1:142–3.
345. Little, "The Jews in Christian Europe," 288.
346. Thomas of Cantimpré, *Bonum universale de apibus* 3.29.23; and Keller, *Diaspora*, 223.
347. Grayzel, *The Church and the Jews*, 10, n7.
348. Roth, "The Feast of Purim," 524.
349. Trachtenberg, *The Devil and the Jews*, 6, 31, 50–l, 83, 124–55; Haym Maccoby, *The Sacred Executioner* (New York 1982), 152–60; Keller, *Diaspora*, 223.
350. For Frederick II's Charter of 1236, see Chazan, *Church, State, and Jew*, 123–6. See also Langmuir, *History, Religion, and Antisemitism*, 263–4, 299–300; Grayzel, *The Church and the Jews*, 339–40.
351. Simonsohn, *History*, 50 n30.
352. "Blood Libel," *Encyclopaedia Judaica* (New York 1971), 4:1122.
353. Grayzel, *The Church and the Jews*, 73.
354. See Trachtenberg, *The Devil and the Jews*, 135, 166.
355. Poliakov, *The History of Antisemitism*, 1:148.
356. Simonsohn, *History*, 83; R. Po-chia Hsia, *The Myth of Ritual Murder: Jews and Magic in Reformation Germany* (New Haven 1988), 43–5.
357. The anti-Jewish German play, *Das Endenger Judenspiel*, recapitulates the Simon of Trent ritual-murder defamation. Lazar, "The Lamb and the Scapegoat," 58.
358. Hsia, *The Myth of Ritual Murder*, 43–50.
359. Simonsohn, *History*, 85; Poliakov, *The History of Antisemitism*, 1:148.
360. Anton Bonfin, *Rerum Hungaricum Decades*, Dec. 5, Book 4.
361. Jean Stengers, *Les Juifs dans les Pays-Bas au Moyen Age* (Brussels 1950), 55–6.
362. Hsia, *The Myth of Ritual Murder*, 223.
363. Hsia, *The Myth of Ritual Murder*, 30, 222.
364. Chrysostom, "Homily on Matthew", 33, 40.
365. Maccoby, *The Sacred Executioner*", 159–61.
366. Hsia, *The Myth of Ritual Murder*, 55.
367. Marcus, *The Jew in the Medieval World*, 155.
368. Maccoby, *The Sacred Executioner*, 159.
369. Trachtenberg, *The Devil and the Jews*, 117.
370. Simonsohn I, doc. 275.
371. Chazan, *Medieval Jewry in Northern France*, 180–2.
372. Little, "The Jews in Christian Europe," 288; Lazar, "The Lamb and the Scapegoat," 57.

373. Langmuir, *History, Religion, and Antisemitism*, 261.
374. O.B. Hardison, *Christian Rite and Christian Drama in the Middle Ages* (Baltimore 1965), 135.
375. Simonsohn I, doc. 82; Grayzel I, doc. 18.
376. Maccoby, *The Sacred Executioner*, 159.
377. Chazan, *Medieval Jewry in Northern France*, 180–2.
378. Mentioned by Josef ha-Cohen in *Emek ha-bacha, Vallée des Pleurs* [Vale of Tears] (Paris 1881), 28.
379. Cohen, *The Friars and the Jews*, 239.
380. Léon Poliakov, *Histoire de l'Antisémitisme* (Paris 1965), 1:99–100, 316 n131; Nicholas de Lange, *Atlas of the Jewish World* (New York 1984), 36.
381. Keller, *Diaspora*, 235–6, 230–1; Poliakov, *The History of Antisemitism*, 1:99–100; Graetz, *The History of the Jews*, 4:35–7; Salo Baron, *Social and Religious History of the Jews* (Philadelphia 1952–69), 11:265–6.
382. Edward Flannery, *The Anguish of the Jews* (New York 1985), 108–9.
383. Flannery, *The Anguish of the Jews*, 108–9; Graetz, *The History of the Jews*, 4:97–8; Baron, *Social and Religious History of the Jews* 11:416–17.
384. *A Horrible Thing Which Was Done at Passau by the Jews . . .*, Marcus, *The Jew in the Medieval World*, 156–7.
385. Hsia, *The Myth of Ritual Murder*, 50–6.
386. William Monter, *Ritual, Myth, and Magic in Early Modern Europe* (Brighton 1983), 18; Paul Grosser and Edwin Halperin, *Antisemitism: The Causes and Effects of Prejudice* (Secaucus 1979), 126–31.
387. Monter, *Ritual, Myth, and Magic*, 18.
388. Simonsohn, *History*, 76–7 n97.
389. Simonsohn, *History*, 76; Simonsohn II, docs. 805, 812.
390. Hsia, "Jews as Magicians in Reformation Germany," 122.
391. Peter Browe, "Die Hostienschändungen der Juden im Mittelalter," *Römische Quartalschrift für christlische Altertumskunde und für Kirchengeschichte* 34 (1926), 197. See Will Erik Peukert, "Ritualmord," in *Handwörterbuch des deutschen Aberglaubens* (Berline 1935–36), 7:727–39.
392. Dennis Showalter, *Little Man, What Now? Der Stürmer in the Weimar Republic* (Hamden 1982), 106.
393. *Codex Theodosianus*, 16:5:44.
394. The Council decreed that Jewish parents who circumcised their children were to have their property confiscated and their noses cut off.
395. Grayzel I, 109.
396. Grayzel I, 333.
397. Revel-Neher, *The Image of the Jew in Byzantine Art*, 35.
398. Séraphine Guerchberg, "The Controversy over the Alleged Sowers of the Black Death in the Contemporary Treatises on Plague," in Sylvia Thrupp, ed., *Change in Medieval Society* (New York 1964), 209.
399. A medieval Christian movement taking its name from the Italian *flagello*, or whip. Flagellants believed that Christian sin caused the plague and other calamities; as a result, they appeared to whip themselves to pay God the penance they imagined God required to lift the punishment. But such

activity often degenerated into lawlessness that resulted in the slaughter of Jews.

400. In this instance, the surviving Jews were protected by King Peter IV of Aragon. "Letters to the Councilors and Citizens of Barcelona" (24 May and 3 June 1348 and 26 February 1349), in Chazan, *Church, State, and Jew*, 128–31.
401. Keller, *Diaspora*, 237; see also Cohn, *Pursuit of the Millennium*, chapter 7.
402. Grosser and Halperin, *Antisemitism*, 129–31.
403. Jacob von Königshofen, in Marcus, *The Jew in the Medieval World*, doc. 9ii.
404. Graetz, *The History of the Jews*, 4:108.
405. Philip Ziegler, *The Black Death* (New York 1969), 98, 100–2.
406. Ziegler, *The Black Death*, 103. See also Friedrich Closener, *Strassburger Deutsche Chronik*, in J. Höxter, *Quellenbuch zur jüdischen Geschichte und Literatur* 3 (1927), 28–30.
407. Jean de Preis (d'Outremeuse), "Ly Myreur des Histors," in Stanislas Bormans, ed., *Chroniques belges* (Brussels 1880), 6:387.
408. Quoted by Roskies, *Against the Apocalypse*, 46.
409. Mordecai Breuer, "The 'Black Death' and Antisemitism," in Almog, *Antisemitism through the Ages*, 144–6.
410. Fritz Baer, in *Zion* 3 (1938), in Alex Bein, *The Jewish Question* (New York 1990), 507.
411. John Edwards, "Mission and Inquisition among *Conversos* and *Moriscos* in Spain, 1250–1550," in Sheils, *Persecution and Toleration*, 140. See also Peter Browe, *Die Judenmission im Mittelalter und die Päpste* (Rome 1942).
412. Simonsohn I, doc. 243.
413. *Codex Theodosianus* 16:8:23, of 24 September 416.
414. Gratian, *Decreti*, I, XLV, 5.
415. Grayzel, "The Papal Bull *Sicut Judeis*," in Cohen, *Essential Papers*, 249.
416. Gratian's *Concordantia Discordantium Canonum*, or *Decretum Gratiani*. Cross, *The Oxford Dictionary of the Christian Church*, 589.
417. Simonsohn, *History*, 258.
418. Simonsohn I, doc. 77; Grayzel I, doc. 12.
419. Grayzel II, doc. 53.
420. Letters of Pope Innocent VI to the doge of Venice in 1356. Simonsohn I, docs. 379–81.
421. Aquinas, *Summa Theologiae*, Book 2, Part 2, Question 10, Article 8.
422. Aquinas, *Summa Theologiae*, Book 2, Part 2, Question 10, Article 8.
423. Aquinas, *Summa Theologiae*, Part III, Question 68, Article 10; Book 2, Part 2, Question 10, Article 10.
424. Pakter, *De His Qui Foris Sunt*, 298–9.
425. Simonsohn, *History*, 71, 255.
426. John Duns Scotus, *IV Sententiarum*, d. 4, q. 9, no. 1, in *Opera Omnia* (Paris 1894), 16:487–8.
427. Yerushalmi, "Response to Rosemary Ruether," 100. See also Simonsohn, *History*, 29.
428. Sophia Menache, *The Vox Dei* (New York 1990), 25.
429. Habermann, *Sefer Gezerot*, 301.

430. William of Newburgh, *Historia Rerum Anglicanarum*, in Jacobs, *Jews of Angevin England*, 123–4. See also Dobson, *The Jews of Medieval York*, 26–7.
431. Simonsohn, *History*, 72–3.
432. Jacobs, *Jews of Angevin England*, 215. For a charter very favorable to the Jews, see *The Charter of the Jews of the Duchy of Austria, 1 July 1244* in Marcus, *The Jew in the Medieval World*, doc. 6. For Eugenius IV, see the papal letters *Super gregem Dominicum* and *Venerabilibus fratribus*, in Simonsohn II, docs. 740, 741.
433. H.H. Ben-Sasson, "Popular Pressure against the Status of the Jews," in H.H. Ben-Sasson, ed., *A History of the Jewish People* (London 1976), 578–80.
434. Cecil Roth, *The Jews in the Renaissance* (Philadelphia 1959), 14.
435. See Heath Dillard, *Daughters of the Reconquest: Women in Castilian Town Society, 1100–1300* (Cambridge, Eng. 1989), 152, 206–7.
436. Simonsohn, *History*, 366; Manuel, *Broken Shaft*, 23.
437. This description is contained in a letter from Pope Clement VI in 1346, offering a dispensation to a priest who had persecuted these Jews. Simonsohn I, doc. 369.
438. Every medieval persecution of Jews resulted in attacks on synagogues. Simonsohn, *History*, 124.
439. Roth, *A History of the Marranos*, 14–17.
440. Glatzer, "Pablo de Santa Maria on the Events of 1391," 130, 135.
441. Glatzer, "Pablo de Santa Maria on the Events of 1391," 130, 135; Manuel, *Broken Shaft*, 24–5.
442. Baer, *The Jews in Christian Spain*, 2:95–9, 125, 424; Keller, *Diaspora*, 247, 250–1.
443. Raymond Carr, "Corruption or Expulsion," Review of Elie Kedourie, ed., *Spain and the Jews* (London 1992), in *The Spectator* (22 May 1992) 30; Edwards, "Mission and Inquisition," 146–7.
444. S. Mitrani-Samarian, "A Valencian Sermon of Saint Vincent Ferrer," *Revue des études juives* 54 (1895), 241.
445. Baer, *The Jews in Christian Spain*, 2:166–9.
446. Kenneth Stow, "The Burning of the Talmud," in Cohen, *Essential Papers*, 402. Less than 40 years after he died, Ferrer was canonized. Simonsohn II, 844.
447. Simonsohn, *History*, 320.
448. Simonsohn, *History*, 320.
449. Simonsohn II, doc. 522.
450. Simonsohn II, docs. 530, 532–5, 537–52, 554–5, 558.
451. *Etsi doctoris* (11 May 1415), in Simonsohn II, doc. 538. Simonsohn, *History*, 322 n75.
452. Baer, *The Jews in Christian Spain*, 179.
453. Simonsohn III, doc. 1000; Simonsohn III, docs. 1024a, 1032–4.
454. Simonsohn III, docs. 1000 and 1040.
455. Roth, *A History of the Marranos*, 21, 29–30.
456. *Historia de los Reyes Católicos*, book I, chapter 44, 600, in Baer, *The Jews in Christian Spain*, 2:327–8.
457. Roth, *A History of the Marranos*, 21, 29–30.

458. Lazar, "The Lamb and the Scapegoat," 58; Roth, *A History of the Marranos*, 52; Henry Charles Lea, "Santo Niño de la Guardia," *English Historical Review* 4 (1889), 229–50.
459. Simonsohn III, doc. 1158a.
460. Keller, *Diaspora*, 261; see also Simonsohn, *History*, 379.
461. Maurice Kriegel, "La jurisdiction inquisitoriale sur les Juifs à l'époque de Philippe le Hardi et Philippe le Bel," in Myriam Yardeni, ed., *Les Juifs dans l'histoire de France* (Leiden 1980), 77.
462. Simonsohn III, doc. 1158a.
463. Jonathan Israel, *European Jewry in the Age of Mercantilism* (Oxford 1989), 23.
464. Moslems who refused baptism were expelled from Castile in 1502 and from Aragon in 1610.
465. Johannes Eck, *Ains Judenbüechlins Verlegung* (Ingolstadt 1541), in Hsia, *The Myth of Ritual Murder*, 130.
466. Pico della Mirandola, "Adversus astrologis," V, chapter 1, xii, in *Opera* (1504).
467. His brother, Luigi, blamed *conversos* in the Spanish Army for the 1527 Sack of Rome. Simonsohn, *History*, 89 n134.
468. Francesco Guicciardini, "Relazione de Spagna," in *Opere* (Milan 1956), 34–5.
469. Pierre Bayle, *Pensées diverses sur la Comète and Commentaire philosophique, in Oeuvres diverses* (The Hague 1727), 3:125a, 2:419b–420a.
470. "Germany and the Jews: Two Views," *Conservative Judaism* (Fall 1962/Winter 1963), 43.

CHAPTER 4 THE GERMANIES FROM LUTHER TO HITLER

1. See Hsia, *The Myth of Ritual Murder*, 112–16, 116–18; Shlomo Simonsohn, *The Apostolic See and the Jews: Documents, 492–1404* (Toronto1988), [Simonsohn I], 257 n61.
2. Oberman, *The Roots of Antisemitism*, 49; Hsia, "Jews as Magicians in Reformation Germany," 125.
3. Oberman, *The Roots of Antisemitism*, 26, 41–3, 50, 78, 95. See also Christoph Peter Burger, "Endzeiterwartung im späten Mittelalter," in *Der Antichrist und Die Fünfzehn Zeichen vor dem Jungsten Gericht* (Hamburg 1979), 43.
4. Luther had been an Augustinian friar.
5. Quoted by J. Janssen, *Die allgemeinen Zustände des deutschen Volkes beim Ausgang des Mittelalters* (Freiburg 1887), 1:399.
6. Johannes von Staupitz, *Salzburger Passionspredigten*, 1512, Sermon X, in Staupitz, *Sämtliche Schriften*, ed. Graf zu Dohna von Lothar and R. Wetzel (Berlin 1983).
7. Eck, *Ains Judenbüechlins Verlegung*, and *Christliche Auslegung der Evangelien von der Zeit . . .* (Tubingen 1531), in Hsia, *The Myth of Ritual Murder*, 127, 130.

8. Manuel, *The Broken Shaft*, 49.
9. Salo Baron, "John Calvin and the Jews," in Cohen, *Essential Papers*, 381.
10. Baron, "John Calvin and the Jews," 384. Servetus was a physician and heretical theologian from Navarre who denied the Trinity, thus offending both Catholics and Protestants. Having escaped from the Inquisition, he was burned at the stake as a heretic by Calvin.
11. Manuel, *Broken Shaft*, 50.
12. Jacques Courvoisier, "Calvin et les Juifs," *Judaica, Beiträge zum Verstandnis des Jüdischen Schicksals in Vergangenheit und Gegenwart* (Zurich 1946), 206.
13. Francesco Guicciardini, "Relazione de Spagna," in *La legazine di Spagna, ossia carteggio tenuta da Guicciardini, ambasciatore della Repubblica fiorentina presso Ferdinanco il Cattolico, 1512–1513* in *Opere* (Florence 1864), 6:270–2.
14. David Ruderman, *The World of the Renaissance Jew: The Life and Thought of Abraham ben Mordecai Farissol* (Cincinnati 1981), 101–2, 105–6.
15. John Kleiner, "The Attitudes of Martin Bucer and Landgrave Philipp toward the Jews of Hesse," in Richard Libowitz, ed., *Faith and Freedom* (Oxford 1987), 221–30. See also Hastings Eells, "Bucer's Plan for the Jews," *Church History* 6 (1937), 127–35.
16. Hsia, "Jews as Magicians in Reformation Germany," 119, 122, 124–5.
17. Erasmus, Letter to Jakob von Hochstraten, 11 August 1519.
18. Erasmus, Letter to Wolfgang Capito, 26 February 1517.
19. Martin Stöhr, "Martin Luther und die Juden," in Heinz Kremers et al., eds., *Die Juden und Martin Luther: Martin Luther und die Juden: Geschichte, Wirkungsgeschichte, Herausforderung* (Darmstadt 1987), 90. See also Heiko Oberman, *Luther: Between Man and Devil* (New Haven 1989).
20. See Hsia, "Jews as Magicians in Reformation Germany," 124; Manuel, *Broken Shaft*, 47.
21. In *Vom Schem Hamphoras und das Geschlect Christi* (*On the Holy Name and the Lineage of Christ*).
22. Gordon Rupp, "Luther against 'the Turk, the Pope, and the Devil,' " in Peter Brooks, ed., *Seven-Headed Luther: Essays in Commemoration of a Quincentenary, 1483–1983* (Oxford 1983), 258–9. Neither Rupp nor the other essayists in this collection deal with the Jews.
23. Johannes Brosseder, "Luther und der Leidensweg der Juden," in Kremers et al., *Die Juden und Martin Luther*, 119; Adam Weyer, "Die Juden in den Predigten Martin Luthers," in Kremers et al., *Die Juden und Martin Luther*, 162–70; Heiko Oberman, "Die Juden in Luthers Sicht," in Kremers et al., *Die Juden und Martin Luther*, 145–7; Ben-Zion Degani, "Die Formulierung und Propagierung des jüdischen Stereotypes in der Zeit vor der Reformation und sein Einflus auf den jungen Luther," in Kremers et al., *Die Juden und Martin Luther*, 37.
24. See, for example, his sermons on the *Book of Psalms*, in *D. Martin Luther's Werke: Kritische Gesamtausgabe* (Weimar 1883–), 56: 46, 199 (hereafter cited as WA).
25. "Martin Luther's to George Spalatin," from Luthers Correspondence and Other Contemporan, Letters, trans. P. Smith (1913), Vol. 1, 28–9; *Modern History Sourcebook*, http://www.fordham.edu/halsall/mod/1514luther.html

26. Luther, "That Jesus Christ Was Born a Jew," 45:199–229, Jaroslav Pelikan and Helmut Lehmann, general eds. *Dass Jesus Christus ein geborener Jude sei* may be found in the Weimar edition, 11:314–36; *Von den Juden und ihren Lügen* in 53:417–52; *Vom Schem Hamphoras und vom Geschlecht Christi* in 53:579–648; and Luther's last sermon at Eisleben in 51:187–96; the sermon's final section, *Vermahnung wider die Juden* in 51:195–6.
27. Luther, "That Jesus Christ Was Born a Jew," 200–1, 229.
28. See Lewin, *Luthers Stellung*; Wilhelm Maurer, *Kirche und Synagogue: Handbuch zur Geschichte von Christen und Juden* (Stuttgart 1968), 1:375.
29. Luther, "That Jesus Christ Was Born a Jew," 229.
30. *Von die Juden und Ihren Lügen (On the Jews and Their Lies)* and *Vom Schem Hamphoras und das Geschlect Christi (On the Holy Name and the Lineage of Christ).*
31. Simonsohn, *History*, 337 n110; Selma Stern, *Josel of Rosheim* (Philadelphia 1965).
32. Marcus, *The Jew in the Medieval World*, 198.
33. Luther, "Jews," 47:137.
34. Luther, "Jews," 229, 154, 167; see also Luther's Lectures on Genesis of 1538–39 in *Luther's Works* 3:113.
35. *The Merchant of Venice*, Act 2, Scene 4.
36. Luther, "Jews," 160, 164, 299, 215, 167, 213, 154, 289, 141, 217, 277. For other references to Jews as devils, see 139, 214, 226, 228, 232, 241–2, 256, 266, 275.
37. Luther, "Jews," 242, 267, 217.
38. Luther, "Jews," 288, 264–5.
39. Hsia, "Jews as Magicians in Reformation Germany," 121.
40. Shachar, *The Judensau*, 57–8.
41. Shachar, *The Judensau*, 57–8.
42. Isaiah Trunk, *Jewish Responses to Nazi Persecution* (New York 1979), 273.
43. Luther, "Jews," 277, 288, 264–5, 156–7, 164, 256, 212.
44. Obermann, *Luthers Werke* (Erlangen 1854), 32:298, 282; in Hartmann Grisar, *Luther* (St. Louis 1915), 4:286, and 5:406. See also the Weimar edition of Luther's works, 53:587, 600–1, 619.
45. WA 53:636–7.
46. Luther, "Jews," 278.
47. Hans Hillerbrand, "Martin Luther and the Jews," in James Charlesworth, ed., *Jews and Christians* (New York 1990), 132.
48. Luther implies that this is impossible.
49. Luther, "Jews," 269–72.
50. Aquinas, *Opuscula Omnia* (Paris 1927), 1:490. See also Chrysostom, *Homilies against the Jews*, 1.2.4–6, *Demonstration to the Jews and Gentiles That Christ Is God*, 4 in *PG* 48:819; Thomas of Chobham, *Summa confessorum* (Louvain 1968), Question XI, chapter 1, 505, in Le Goff, *Your Money or Your Life*, 42; Aquinas, *Summa Theologiae*, 2a 2ae, 10, 10–12; Baer, *The Jews in Christian Spain*, 2:357.
51. Luther, "Jews," 265.
52. Peter Wiener, *Martin Luther: Hitler's Spiritual Ancestor* (London 1945), 61–2.
53. Luther, "Jews," 272, 292.
54. Obermann, Luther, *Man between God and the Devil*, 289.

55. Luther, "Admonition: Against the Murderous and Thieving Hordes of Peasants," in Roland Bainton, *Here I Stand* (New York 1978), 217.
56. Quoted in Lewin, *Luthers Stellung*, 77. Another advocate of permanently shutting Jewish mouths was John Chrysostom. *Homilies against the Jews*, 5.1.6; 7.2.3.
57. See Alter, "From Myth to Murder," 40.
58. Luther, "Jews," 242, 289, 267.
59. See, for example, Stöhr, "Martin Luther and die Juden," 127–9; and Haim Hillel Ben-Sasson, "Jewish–Christian Disputation in the German Empire," *Harvard Theological Review* 59 (1966), 369–71; Wilhelm Maurer, "Die Zeit der Reformation," in Karl Rengstorf and Siegfried von Kortzfleisch, eds., *Kirche und Synagogue* (Stuttgart 1968), 1:378–80; Lewin, *Luthers Stellung*, 110; Oberman, *The Roots of Antisemitism*, 9.
60. Baron, *A Social and Religious History of the Jews*, 13:429, n26.
61. Israel, *European Jewry in the Age of Mercantilism*, 12–13.
62. Liebe, *Das Judentum in der deutschen Vergangenheit*, 37.
63. Shachar, *The Judensau*, 87–88 n239.
64. Quoted by Bein, *The Jewish Question*, 713.
65. Titled *Kurze Beschreibung und Erzehlung von einem Juden mit Namen Ahasuerus* (*Brief Account and Story of a Jew Called Ahasuerus*). Edelmann, "Ahasuerus the Wandering Jew," 6.
66. Paul Rose, *Revolutionary Antisemitism in Germany from Kant to Wagner* (Princeton 1990), 24.
67. Manuel, *Broken Shaft*, 86–90.
68. Johannes Buxtorf, in Grayzel I, 301 n3.
69. Lange, *Atlas of the Jewish World*, 44; Graetz, *The History of the Jews*, 4:696–8.
70. John Edwards, *The Jews in Christian Europe, 1400–1700* (London 1988), 111–12. For Luther's later influence, see, for example, Uriel Tal, *Christians and Jews in Germany: Religion, Politics, and Ideology in the Second Reich, 1870–1914* (Ithaca 1975); and Richard Gutteridge, *Open Thy Mouth for the Dumb* (Oxford 1976).
71. See K. von der Bach, *Luther als Judenfeind* (Berlin 1931); E. Vogelsang, *Luthers Kampf Gegen die Juden* (Tübingen 1938); and Tal, *Christians and Jews in Germany*.
72. Quoted by Fischer, *The History of an Obsession*, 127.
73. Quoted in Steigmann-Gall, *The Holy Reich*, 24–5, 136.
74. Quoted in ibid., 137.
75. Steigmann-Gall, *The Holy Reich*, 120–1.
76. Steigmann-Gall, *The Holy Reich*, 235.
77. Luther, "Jews," 272, 292.
78. *Trial of Major War Criminals Before the International Military Tribunal* (Nuremberg 1947–49), 12:318.
79. Heiko Obermann, "The Nationalist Conscription of Martin Luther," in Carter Lindberg, ed., *Piety, Politics, and Ethics: Reformation Studies in Honor of George Wolfsans Farell*, Sixteenth Century Essays and Studies, Vol. 3 (Kirkville 1984), 70; John Conway, "Protestant Missions to the Jews, 1810–1980: Ecclesiastical Imperialism or Theological Aberration?" *Holocaust and Genocide Studies* 1 (1986), 135. See also Robert Ericksen, *Theologians Under Hitler* (New Haven 1985).

80. Quoted by Ernst Ehrlich, "Luther und die Juden," in Kremers et al., *Die Juden und Martin Luther*, 86.
81. Wilhelm Niemöller, "The Niemöller Archives," in Franklin Littell and Hubert Locke, eds., *The German Church Struggle and the Holocaust* (Detroit 1974), 54. See also Johan Snoek, *The Grey Book* (New York 1970).
82. Snoek, *The Grey Book*, 291–2.
83. Guido Kisch, "Necrologue Reinhold Lewin, 1888–1942," *Historia Judaica* 8 (1946), 217–19.
84. Lewin, *Luthers Stellung*, 110.
85. Chamberlain, *Deutschlands Kampfziel* (Munich 1916), 61.
86. See Jacob Katz, *From Prejudice to Destruction* (Cambridge, MA 1980), 63.
87. *From Prejudice to Destruction*, 21.
88. Katz, *From Prejudice to Destruction*, 20.
89. Israel, *European Jewry in the Age of Mercantilism*, 234–6; Hsia, "Jews as Magicians in Reformation Germany," 132.
90. Richard Levy, *Antisemitism in the Modern World: An Anthology of Texts* (Lexington, MA 1991), doc. 1.
91. Manuel, *Broken Shaft*, 153.
92. *Grand Complete Universal Lexicon*, published in 1735.
93. In Manuel, *Broken Shaft*, 250–1.
94. "*Diabolus in musica.*" Kurt Pahlen et al., *The World of the Oratorio* (Portland 1985), 28–34.
95. Orlando di Lasso, *St. Matthew Passion* (New York 1967). See also Bein, *The Jewish Question*, 556.
96. Pahlen et al., *The World of the Oratorio*, 23–8. See the English version of Robert Shaw and Henry Drinker, *The Passion According to John* (n.c., n.d.).
97. Leonard Swidler and Gerald Sloyan, *Commentary on the Oberammergau Passionspiel in Regard to Its Image of Jews and Judaism* (New York 1978).
98. Rappaport, "The Ritual Murder Accusation," 322.
99. Manuel, *Broken Shaft*, 249.
100. See Hans Joachim Schoeps, "Philosemitism in the Baroque Period," *Jewish Quarterly Review* 47 (1956–57), 139–44.
101. Alfred Low, *Jews in the Eyes of the Germans* (Philadelphia 1979), 41–7.
102. Low, *Jews in the Eyes of the Germans*, 41.
103. Jacob Katz, *Out of the Ghetto: The Social Background of Jewish Emancipation, 1770–1870* (New York 1978), 47.
104. Manuel, *Broken Shaft*, 275.
105. Selma Stern, *Jud Süss: Ein Betrag zur deutschen und zur jüdischen Geschichte* (Berlin 1929).
106. David Sorkin, *The Transformation of German Jewry, 1780–1840* (New York 1987), 8 and throughout.
107. Lessing, in Bein, *The Jewish Question*, 177, 547, 562.
108. In Low, *Jews in the Eyes of the Germans*, 66.
109. Mayer Kayserling, *Moses Mendelssohn* (Leipzig 1862), 546.
110. In Low, *Jews in the Eyes of the Germans*, 55–63.
111. Rose, *Revolutionary Antisemitism*, 11.

112. Herder, *Ideen zur Philosophie der Geschichte der Menschheit, in Werke,* 14:58–63, in Low, *Jews in the Eyes of the Germans,* 63.
113. Poliakov, *The History of Antisemitism,* 3:17–18.
114. Quoted in Hannah Arendt, *The Origins of Totalitarianism* (Cleveland 1958), 30.
115. Von Dohm, *Über die bürgerliche Verbesserung der Juden, (Concerning the Amelioration of the Civil Status of the Jews)* (Berlin 1781, 1783), in Chazan and Raphael, *Modern Jewish History*; Katz, *Out of the Ghetto,* 50.
116. "*Bürgerliche Gesellshaft.*" See Manuel, *Broken Shaft,* 275.
117. See Manuel, *Broken Shaft,* 276–7; George Mosse, *Germans and Jews* (New York 1971), 39.
118. Dohm, *Über die bürgerliche Verbesserung der Juden,* doc. 1:2–13.
119. Quoted by Keller, *Diaspora,* 392.
120. Chadwick, *The Popes and European Revolution,* 435; Manuel, *Broken Shaft,* 273.
121. Quoted by Poliakov, *The History of Antisemitism,* 3:287.
122. Goethe, Letter to Karl Friedrich Zelter, 19 May 1812, in Bein, *The Jewish Question,* 585.
123. Goethe, in *Goethe's Werke* (Weimar 1887–1918), 52:267–7, 37:59–60; Low, *Jews in the Eyes of the Germans,* 70.
124. *Wilhelm Meisters Wanderjahre* in *Goethe's Werke,* 25:210, in Isaac Barzilay, "The Jew in the Literature of the Enlightenment," *Jewish Social Studies* 18 (October 1956), 246 n10.
125. Quoted in Low, *Jew in the Eyes of the Germans,* 72, 83.
126. Low, *Jews in the Eyes of the Germans,* 71.
127. Quoted in Keller, *Diaspora,* 367.
128. *Religion within the Limits of Reason Alone.* See Low, *Jews in the Eyes of the Germans,* 93–4; Manuel, *Broken Shaft,* 288.
129. Immanuel Kant, *Die Religion innerhalb der Grenzen der blossen Vernunft* in Bein, *The Jewish Question,* 560, 563. See also Sorkin, *The Transformation of German Jewry,* 22; Katz, *From Prejudice to Destruction,* 65–8.
130. Kant, *Anthropologie in pragmatischer Hinsicht* in Bein, *The Jewish Question,* 560.
131. Manuel, *Broken Shaft,* 286.
132. Low, *Jews in the Eyes of the Germans,* 94–5.
133. Manuel, *Broken Shaft,* 290.
134. Poliakov, *The History of Antisemitism,* 3:179, 3:511 n45.
135. Johann Fichte, *Beitrag zur Berichtigung der Urteile des Publikums über die französische Revolution,* in *Fichtes Sämtliche Werke* (Berlin 1845–46), 6:149–50.
136. Fichte, *Beitrag zur Berichtigung,* 6:149.
137. Fichte, *Beitrag zur Berichtigung,* 6:150.
138. Fichte, *Beitrag zur Berichtigung,* 6:149–50.
139. Fichte, *Reden an die deutsche Nation* (1808), Sixth Address, Point 81.
140. Eleonore Sterling, *Judenhass: Die Anfänge des politischen Antisemitismus in Deutschland, 1815–1850* (Frankfurt 1969), 128–9.
141. See Tal, *Christians and Jews in Germany,* 291–2; George Mosse, *Toward the Final Solution: A History of European Racism* (New York 1978), 119, 121, 127, and chapter 11.

142. Katz, *Out of the Ghetto*, 87.
143. Sterling, *Judenhass*, 51.
144. Sterling, *Judenhass*, 51.
145. Sterling, *Judenhass*; Alter, "From Myth to Murder," 34.
146. Sterling, *Judenhass*, 67–8.
147. Jacob Katz, *Jews and Freemasons in Europe, 1723–1939* (Cambridge, MA 1970), 209.
148. Low, *Jews in the Eyes of the Germans*, 162.
149. Robert Wistrich, *Socialism and the Jews: The Dilemmas of Antisemitism in Germany and Austria–Hungary* (London 1982), 44–5.
150. See Tal, *Christians and Jews in Germany*, 147–8.
151. See Jacob Katz, "Misreadings of Antisemitism," *Commentary* (July 1983), 39–44.
152. Michael Meyer, "The German Jews: Some Perspectives on Their History," in Michael Meyer, ed., *Judaism within Modernity: Essays on Jewish History and Religion* (Detroit 2001), 102.
153. Meyer, "The German Jews," 102.
154. Rose, *Revolutionary Antisemitism*, 29–30.
155. Meyer, "The German Jews," 101, 104.
156. Fischer, *The History of an Obsession*, 70.
157. Meyer, "The German Jews," 104–5.
158. Rose, *Revolutionary Antisemitism*, 263. See also Langmuir, *History, Religion, and Antisemitism*, 306, 310, 313–14, 320; Sterling, *Judenhass*, 13, 167; Bein, *The Jewish Question*, 496.
159. Rose, *Revolutionary Antisemitism*, 139.
160. Low, *Jews in the Eyes of the Germans*, 121, 123.
161. Peter Pulzer, *The Rise of Political Antisemitism in Germany and Austria* (Cambridge, MA 1988), 32n.
162. Katz, *Out of the Ghetto*, 76, 102–3.
163. Ibid., 228 n18.
164. H.W. Koch, *A History of Prussia* (London 1987), 179–80.
165. Quoted in Poliakov, *The History of Antisemitism*, 3:292.
166. Koch, *History of Prussia*, 180.
167. Low, *Jews in the Eyes of the Germans*, 123–33.
168. See Waagenaar, *The Pope's Jews*, 253–4.
169. See Tal, *Christians and Jews in Germany*, 101 n57.
170. Quoted in Kertzer, *Popes against the Jews*, 126–7.
171. Michael Burns, *Dreyfus: A Family Affair, 1789–1945* (New York 1991), 17, 21, 31.
172. Keller, *Diaspora*, 390, 392–3; Katz, *From Prejudice to Destruction*, 100.
173. Sterling, *Judenhass*, 157–9, 162–3, 171.
174. Low, *Jews in the Eyes of the Germans*, 115.
175. Low, *Jews in the Eyes of the Germans*, 31–4.
176. Boerne, *Briefe aus Paris*, Letter 74, in Arendt, *The Origins of Totalitarianism*, 64.
177. Low, *Jews in the Eyes of the Germans*, 134.
178. In Katz, *Out of the Ghetto*, 78.

179. Low, *Jews in the Eyes of the Germans*, 138.
180. Quoted in Poliakov, *The History of Antisemitism*, 3:294.
181. *Wilhelm und Caroline von Humboldt in ihren Briefen* (Berlin 1900), 5:236. See also Hannah Arendt, "Privileged Jews," *Jewish Social Studies* 8 (1946), 23.
182. Karl Cerff, ed., *Kulturpolitisches Mitteilungsbelass der Reichspropagandaleitung der NSDAP* (Berlin 1941), 7, in Victor Farías, *Heidegger and Nazism* (Philadelphia 1989), 106; Low, *Jews in the Eyes of the Germans*, 140–1.
183. Rose, *Revolutionary Antisemitism*, 26; Poliakov, *The History of Antisemitism*, 3:293.
184. Lotte Brank Philip, "The Prado Epiphany by Jerome Bosch," in James Snyder, ed., *Bosch in Perspective* (Englewood Cliffs, NJ 1973) 193.
185. Low, *Jews in the Eyes of the Germans*, 185.
186. Low, *Jews in the Eyes of the Germans*, 182–3.
187. Poliakov, *The History of Antisemitism*, 3:186.
188. Poliakov, *The History of Antisemitism*, 3:182. See also Low, *Jews in the Eyes of the Germans*, 279–80.
189. Poliakov, *The History of Antisemitism*, 3:182–3.
190. Katz, *From Prejudice to Destruction*, 68–72; Low, *Jews in the Eyes of the Germans*, 274–5.
191. Quoted in Poliakov, *The History of Antisemitism*, 3:184.
192. *Philosophy of Right* (1820), Section 270.
193. Radical editor of the *German Yearbooks.*
194. Ruge, *Hallische Jahrbücher* II (1839), 1598, in Sterling, *Judenhass*, 101.
195. Nathaniel Hawthorne, "The Marble Faun," in *The Complete Novels and Selected Tales* (New York 1937), 813; Adolf Hitler, *Mein Kampf*, trans. Ralph Mannheim (Boston, MA 1943), 57, 324.
196. Rose, *Revolutionary Antisemitism*, 252–5.
197. Low, *Jews in the Eyes of the Germans*, 283–5.
198. Low, *Jews in the Eyes of the Germans*, 323.
199. Schopenhauer, *Sämtliche Werke* (Wiesbaden 1949–66), 2:279–81, in Katz, *From Prejudice to Destruction*, 74.
200. Low, *Jews in the Eyes of the Germans*, 323.
201. Arthur Schopenhauer, *Parerga and Paralipomena: Short Philosophical Essays* (Oxford 1974), 2:261, 264; quoted by Adolf Leschnitzer, "The Wandering Jew: The Alienation of the Jewish Image in Christian Consciousness," in Hasan-Rokem and Dundes, *The Wandering Jew*, 227–8.
202. Fritz Stern, "The Burden of Success: Reflections on German Jewry," in *Dreams and Delusions* (New York 1987), 111n.
203. Rose, *Revolutionary Antisemitism*, 135.
204. Die Cavallerotti, "Felix Mendelssohn Bartholdy—eine Geschichte kulturellen Antisemitismus in Deutschland im 19. u. 20. Jhrdt," das KulturNetzWerk e. V.—11/02, <http://www.cavallerotti.de/html/projekt_set/projekte/ mendelssohn/mendelssohn_essay.html>.
205. Quoted by Fischer, *The History of an Obsession*, 101.
206. Rose, *Revolutionary Antisemitism*, 135; Sterling, *Judenhass*, 56.
207. Rose, *Revolutionary Antisemitism*, 159.
208. Heinrich Heine, quoted by Katz, *Out of the Ghetto*, 210.

209. Quoted in Edmund Silberner, "Proudhon's Judaeophobia," *Historia Judaica* (April 1948), 74, and "Two Studies of Modern Antisemitism," *Historia Judaica* (October 1952), 101. Bakunin quoted in L.J. Rather, *Reading Wagner: A Study in the History of Ideas* (Baton Rouge 1990), 177–8.
210. George Lichtheim, "Socialism and the Jews," *Dissent* (July–August 1968), 314–42.
211. Poliakov, *The History of Antisemitism*, 3:417.
212. Michael Meyer, *Jewish Identity in the Modern World* (Seattle 1990), 140–2.
213. Low, *Jews in the Eyes of the Germans*, 261, 263, 267–8.
214. Rose, *Revolutionary Antisemitism*, 29–30, 171.
215. Low, *Jews in the Eyes of Germans*, 162.
216. H.D. Schmidt, "The Terms of Emancipation, 1781–1812," *Leo Baeck Institute Yearbook* I (1956), 36. See also Reinhard Rürup, *Emanzipation und Antisemitismus* (Göttingen 1975).
217. Low, *Jews in the Eyes of the Germans*, 302.
218. See Sterling, *Judenhass*, 139–47, 162–4, 171–4.
219. Quoted by Sterling, *Judenhass*, 141.
220. Sterling, *Judenhass*, 134. See de Lange, *Atlas of the Jewish World*, 57.
221. Wistrich, *Socialism and the Jews*, 56–7, 61.
222. Trachtenberg, *The Devil and the Jews*, chapter 13. See also Steven Aschheim, "The Myth of 'Judaization' in Germany," in Jehuda Reinharz and Walter Schatzberg, eds., *The Jewish Response to German Culture* (Hanover, NH 1985), 215.
223. Aschheim, "The Myth of 'Judaization' in Germany," 220–1.
224. *Atlas of Modern Jewish History* (Oxford 1991).
225. See Bernard Lazare, "La conception sociale du judaïsm et du people juif," and *Le Fumier de Job*, in Michael Marrus, *The Politics of Assimilation: A Study of the French Jewish Community at the Time of the Dreyfus Affair* (Oxford 1971), 190–2.
226. Rose, *Revolutionary Antisemitism*, 263–72.
227. Rose, *Revolutionary Antisemitism*, 263–72.
228. See Rose, *Revolutionary Antisemitism*, 305.
229. Quoted in Silberner, "Proudhon's Judaeophobia," 74, and "Two Studies of Modern Antisemitism," 101.
230. Poliakov, *The History of Antisemitism*, 3:427.
231. Quoted in Silberner, "Proudhon's Judaeophobia," 74; Poliakov, *The History of Antisemitism*, 3:425.
232. Quoted in Silberner, "Two Studies of Modern Antisemitism," 101; Poliakov, *The History of Antisemitism*, 3:425; Wistrich, *Socialism and the Jews*, 31–2, 46–7.
233. Marx, "On the Jewish Question," in Marx and Engels, *Collected Works* (New York 1975–), 3:170–1.
234. Hermann Rauschning, *Hitler Speaks* (London 1919), 235–6.
235. Paul Mendes-Flohr and Jehuda Reinharz, eds., *The Jew in the Modern World: A Documentary History* (Oxford 1980), chapter 7, doc. 10.

236. See Tal, *Christians and Jews in Germany*, chapter 3; Katz, *From Prejudice to Destruction*, chapter 15.
237. Tal, *Christians and Jews in Germany*, 148.
238. Katz, *From Prejudice to Destruction*, 252–3.
239. Katz, *From Prejudice to Destrudction*, 252–4.
240. Tal, *Christians and Jews in Germany*, 148.
241. See Tal, *Christians and Jews in Germany*, 248–59.
242. Pulzer, *The Rise of Political Antisemitism*, 95, 274; Wistrich, *Socialism and the Jews*, 61, 103.
243. Pulzer, *The Rise of Political Antisemitism*, 88.
244. Wistrich, *Socialism and the Jews.*
245. Levy, *Antisemitism in the Modern World*, doc. 4, 58–66.
246. Levy, *Antisemitism in the Modern World*, doc. 4, 58–66.
247. Rose, *Revolutionary Antisemitism*, 287.
248. Rose, *Revolutionary Antisemitism*, 287.
249. Tal, *Christians and Jews in Germany*, 264.
250. Zimmermann, *Wilhelm Marr*, 83, 88–94, 105, 107, 112; Rose, *Revolutionary Antisemitism*, 14; Tal, *Christians and Jews in Germany*, 264; Manuel, *Broken Shaft*, 306.
251. Poliakov, *The History of Antisemitism*, 3:298; Pulzer, *The Rise of Political Antisemitism*, 91–2.
252. Levy, *Antisemitism in the Modern World*, doc. 10, 125–7; Pulzer, *The Rise of Political Antisemitism*, 266–7.
253. Mosse, *Toward the Final Solution*, 139.
254. Rose, *Revolutionary Antisemitism*, 241; Michael Burleigh and Wolfgang Wippermann, *The Racial State: Germany, 1933–1945* (Cambridge, Eng. 1991), 27.
255. Treitschke, *History of Germany* (New York 1919), 7:452.
256. See Katz, *From Prejudice to Destruction*; Tal, *Christians and Jews in Germany*, 55; and Jack Wertheimer, *Unwelcome Strangers: East European Jews in Imperial Germany* (New York 1987).
257. Treitschke, *Ein Wort über unser Judentum*, in Mendes-Flohr and Reinharz, *Jew in the Modern World*, 280–3.
258. Robert Lougee, *Paul de Lagarde, 1827–1891: A Study of Radical Conservatism in Germany* (Cambridge, MA1961), 151–4.
259. Cohn, *Warrant for Genocide*, 171; Pulzer, *The Rise of Political Antisemitism*, 79–80. See Luther, "The Jews and Their Lies," 137–8, 265.
260. In Pulzer, *The Rise of Political Antisemitism*, 80.
261. Lougee, *Paul de Lagarde* 135, 139.
262. *Juden und Indogermanen* (Göttinger 1887), 339, in Alex Bein, "Modern Antisemitism and Its Effect on the Jewish Question," *Yad Vashem Studies* 3 (1959), 14.
263. Quoted by Fritz Stern, *The Politics of Cultural Despair: A Study in the Rise of the Germanic Ideology* (Berkeley 1974), 61–2.
264. Stern, *The Politics of Cultural Despair*, 63n.

265. In two major works, *Ahasverus oder die Judenfrage* (Berlin 1844) and *Der Föderalismus* (Mainz 1879).
266. In Rose, *Revolutionary Antisemitism*, 344–5.
267. See Pulzer, *The Rise of Political Antisemitism*, 73; Rose, *Revolutionary Antisemitism*, 345–6.
268. Rose, *Revolutionary Antisemitism*, 353–4.
269. Tal, *Christians and Jews in Germany*, 139–40.
270. Prof. A. Suchsland, Chair of the Conservative Society of Halle, in Tal, *Christians and Jews in Germany*, 141–2, 315.
271. Charlesworth, *Jews and Christians*, 14; Adolf von Harnack, *What Is Christianity* (New York 1957), 9; Harnack, *Marcion: Das Evangelium vom fremden Gott*, 217, quoted by Rather, *Reading Wagner*, 125. See also Tal, *Christians and Jews in Germany*.
272. Quoted by Robert Osborne, "The Christian Blasphemy," in Charlesworth, *Jews and Christians*, 219.
273. For the Jewish response, see Tal, *Christians, and Jews in Germany*, esp. 65.
274. Low, *The Jews in the Eyes of the Germans*, 373–4.
275. Mommsen, *Auch ein Wort über unser Judentum*, in Mendes-Flohr and Reinharz, *Jew in the Modern World*, 284–7; Tal, *Christians and Jews in Germany*, 50–1; Walter Boehlich, *Der Berliner Antisemitismusstreit* (Frankfurt 1965), 219–20.
276. Tal, *Christians and Jews in Germany*, 51–2.
277. *The Speeches of Adolf Hitler*, ed. Norman Baynes (New York 1969), 17, 59–60, 259, 504, 821, 918, 1372.
278. *Hitler's Table Talk, 1941–44* (London 1973), 141.
279. Tal, *Christians and Jews in Germany*, 89–90.
280. Tal, *Christians and Jews in Germany*, 94.
281. Wistrich, *Socialism and the Jews*, 187–8.
282. Langmuir, *History, Religion, and Antisemitism*, 325.
283. George Berkley, *Vienna and Its Jews* (Cambridge, MA 1988), 77–80.
284. Poliakov, *The History of Antisemitism*, 3:456; Berkley, *Vienna and Its Jews*, 77–80.
285. Kertzer, *Popes against the Jews*, 136.
286. Katz, *From Prejudice to Destruction*, 276.
287. Wistrich, *Socialism and the Jews*, 187.
288. See Peter Viereck, *Metapolitics: From the Romantics to Hitler* (New York 1941); Leon Stein, *The Racial Thinking of Richard Wagner* (New York 1950); Otto Dov Kulka, "Richard Wagner und die Anfänge des modernen Antisemitismus," *Leo Baeck Institute Yearbook* 16 (1961); Margaret Brearley, "Hitler and Wagner: The Leader, the Master, and the Jews," *Patterns of Prejudice* (Summer 1988), 3–22; Robert Gutman, *Richard Wagner: The Man, His Mind, and His Music* (New York 1968); Barry Millington, *Wagner* (London 1984); Ernest Newman, *Life of Richard Wagner* (New York 1946, 1968); and Katz, *The Darker Side of Genius.*
289. Katz, *The Darker Side of Genius*, 116–17.
290. See Rather, *Reading Wagner*, 174.

291. Richard Wagner, "Art and Revolution," (1849), trans. William Ashton Ellis, *Richard Wagner's Prose Works* (London 1892–99), 1:153–4 (hereafter cited as *PW*).
292. Wagner, "Herodom and Christendom," *PW*, 6:280–3.
293. See Clement Vollmner, "Richard Wagner and His Jewish Friends," *South Atlantic Quarterly* 37(2) (1937), 214.
294. Cosima Liszt-Bülow was Wagner's mistress from 1864, his wife from 1870.
295. Cosima Wagner, *Die Tagebucher* (Munich 1976), 2:744.
296. Cosima Wagner, *Diaries* (17 December 1881), 2:771.
297. Wagner, "Herodom and Christendom" (1881), *PW*, 6:280.
298. Wagner, "Beethoven" (1870), *PW*, 5:121.
299. Hitler, *Mein Kampf*, 6.
300. Wagner, *Mein Leben* (Munich 1969), 1:27, quoted by Reason, *Reading Wagner*, 115; Wagner, *Mein Leben*, 1:18; Hitler, *Mein Kampf*, 6. Robert Waite reports that although raised a Protestant, Wagner's mother had been a devout Catholic. Waite, *The Psychopathic God: Adolf Hitler* (New York 1977), 127. On Wagner's christology, see Cosima Wagner, *Diaries, 1878–1883* (New York 1980), 1:548 (31 October 1872), 1:691 (28 October 1873), 1:780 (8 August 1874), 1:984 (26 September 1877), 1:1008 (24 December 1877), 2:339 (15 July 1879); Wagner, "Religion and Art" (1880), *PW*, 6:258; George Windell, "Hitler, National Socialism, and Richard Wagner," in De Gaetani, ed., *Penetrating Wagner's Ring*, 223–4; Wagner, "Jesus of Nazareth," *PW*, 8:298–301; Wagner, "Public and Popularity," *PW*, 6:78; Letter to Hans von Wolzogen (17 January 1880) in Wagner, *Selected Letters*, doc. 475; Letter to Ludwig II, 31 March 1881, in Gutman, *Richard Wagner*, 408; Wagner, "Herodom and Christendom," *PW*, 6:283; Wagner, "Religion and Art," *PW*, 6:231; Stefan Kunze, "Parsifal: Das Kunstwerk als sakrale Veranstaltung," in Ersula von Rasuchhaupt and Wolfgang Sedat, eds., *Parsifal: Ein Bühneuweihfestspiel* (Hamburg 1981), 21.
301. Wagner to Liszt, in *Correspondence of Wagner and Liszt* (New York 1897) in Poliakov, *The History of Antisemitism*, 3:439; see also Waite, *The Psychopathic God*, 127–8.
302. Wagner, "Judaism in Music," *PW*, 3:80–2.
303. See "The Artwork of the Future" (1849), *PW*, 1:155. See also, Robert Michael, "Wagner, Christian Anti-Jewishness, and the Jews: A Reexamination," *Patterns of Prejudice* (Fall 1992).
304. Wagner, "Judaism in Music" (1850, 1869), *PW*, 3:100, 120. See also Robert Herzstein, "Richard Wagner at the Crossroads of German Antisemitism, 1848–1933: A Re-Interpretation," *Zeitschrift fur die Geschichte der Juden* 47 (1967), 124–7.
305. Wagner, "Judaism in Music," *PW*, 3:100.
306. Cosima Wagner, *Diaries*, 2:254 (13 January 1879).
307. 19 September 1881, in Newman, *The Life of Richard Wagner*, 4:637. Wagner, "Prelude to Parsifal," *PW*, 8:388–9.
308. Cosima Wagner, *Diaries* (28 April 1880), 2:471; Newman, *The Life of Richard Wagner*, 4:635.

309. Newman, *The Life of Richard Wagner*, 4:637–8.
310. Wagner, "Know Thyself," *PW*, 6:264–5, 271. See also "Judaism in Music," *PW*, 3:111, and "Religion and Art," *PW*, 6:232.
311. See Frank Josserand, "Richard Wagner and German Nationalism," in De Gaetani, *Penetrating Wagner's Ring*, 207; Poliakov, *The History of Antisemitism*, 3:450; Wagner, "Herodom and Christendom," *PW*, 6:280–3.
312. Wagner, "Judaism and Music" *PW*, 3:82.
313. Wagner, "Know Thyself" (1881), *PW*, 6:268–9. See also "What Is German?" (1865), *PW*, 4:158.
314. Wagner, "Jesus of Nazareth," *PW*, 8:300–1.
315. *Das verfluchte Judensgeschmeiss*. Wagner, *Sämtsliche Briefe* (Leipzig 1967–), 1:177, quoted in Rose, *Revolutionary Antisemitism*, 359–60.
316. Wagner, "Jesus of Nazareth" [opera text], (Winter 1848–49), in *PW*, 8:295.
317. Wagner, "Religion and Art," *PW*, 6:233–4. Wagner, "Public and Popularity" (1878), *PW*, 6:77.
318. Wagner, "Jesus of Nazareth," *PW*, 8:303.
319. To Constantin Frantz (14 July 1879), in Richard Wagner, *Selected Letters of Richard Wagner* (New York 1987); ed. and trans. Stewart Spencer and Barry Millington, doc. 471.
320. Wagner, "Judaism in Music," *PW*, 3:100.
321. Wagner, "Judaism in Music," *PW*, 3:121.
322. Wagner, "What Is German," *PW*, 4:166.
323. Wagner, "Know Thyself," *PW*, 6:274.
324. Wagner, "Judaism in Music," *PW*, 3:86–100.
325. Wagner, "Judaism in Music," *PW*, 3:121–2. See also Wagner to Liszt, in *Correspondence of Wagner and Liszt* in Poliakov, *The History of Antisemitism*, 3:435.
326. To King Ludwig II of Bavaria (22 November 1881), in Wagner, *Selected Letters*, doc. 489.
327. Newman, *The Life of Richard Wagner*, 4:639.
328. On 11 and 14 August 1881. Cosima Wagner, *Diaries*, 2:705. See also Katz, *From Prejudice to Destruction*, 275; Hans Rogger, *Jewish Policies and Right-Wing Politics in Imperial Russia* (Berkeley 1986), 26–7, 34–9.
329. Katz, "German Culture and the Jews," 88.
330. Cosima Wagner, *Diaries*, 2:766 (10 December 1881).
331. Cosima Wagner, *Diaries*, 2:770 (16 December 1881).
332. Cosima Wagner, *Diaries*, 2:773 (18 December 1881). See Sorkin, *The Transformation of German Jewry*, 8 and throughout.
333. Harry Zohn, "Fin-de-Siècle Vienna," in Reinharz and Schatzberg, *The Jewish Response to German Culture*, 138–9.
334. See Farías, *Heidegger and Nazism*, 26; Trachtenberg, *The Devil and the Jews*, 42, 108; Pulzer, *The Rise of Political Antisemitism*, 211; and R.A. Kann, *A Study in Austrian Intellectual History* (London 1960), 57, 76–9, 104.
335. Mosse, *Toward the Final Solution*, 137–8, 141.
336. Pulzer, *The Rise of Political Antisemitism*, 161, 175, 211.
337. Henry Cohn, "Theodore Herzl's Conversion to Zionism," *Jewish Social Studies*, 32 (April 1970), 101–10; Victor Conzemius, "l'Antisemitisme

autrichien au XIXième et au XXième siecles," in Nikiprowetzky, *De l'antijudaïsme antique à l'antisémitisme contemporain*, 197; Pulzer, *The Rise of Political Antisemitism*, 176–7, 198; Berkley, *Vienna and Its Jews*, 98.

338. Pulzer, *The Rise of Political Antisemitism*, 200–1.
339. Wistrich, *Socialism and the Jews*, 197.
340. Berkley, *Vienna and Its Jews*, 384; Mosse, *Toward the Final Solution*, 114; Hermann Glaser, *The Cultural Roots of National Socialism* (Austin 1964), 227.
341. Wistrich, *Socialism and the Jews*, 197.
342. Wistrich, *Socialism and the Jews*, 198.
343. See Richard Levy, *The Downfall of the Antisemitic Political Parties in Imperial Germany* (New Haven 1975); Katz, *From Prejudice to Destruction*; Pulzer, *The Rise of Political Antisemitism*; Shulamit Volkov, "Antisemitism as a Cultural Code," *Leo Baeck Institute Year Book* 23 (1978), 25–46.
344. Pulzer, *The Rise of Political Antisemitism*, 282, 291–2, 298, 305.
345. Katz, "German Culture and the Jews," 94.
346. Dieter Hartmann, "Antisemitism and the Appeal of Nazism," *Political Psychology* (December 1984), 636.
347. Stern, "The Burden of Success," 101; Mack Walker, *German Home Towns* (Ithaca, NY 1971), 271.
348. *Der Durchschnittsmensch*.
349. Showalter, *Little Man*, 14.

CHAPTER 5 CHRISTIAN ANTISEMITISM, THE GERMAN PEOPLE, AND ADOLF HITLER

1. Fischer, *The History of an Obsession*, 131.
2. See Langmuir, "Toward a Definition of Antisemitism."
3. In Denmark and Italy, and isolated instances elsewhere, the proportion of Righteous Christians was large. See Hallie, *Lest Innocent Blood Be Shed*; Harold Flender, *Rescue in Denmark* (New York 1963); and Leo Goldberger, ed., *The Rescue of the Danish Jews* (New York 1987).
4. For a Jewish response to this situation, see Anne Roiphe, "Christians and Jews," in *A Season for Healing: Reflections on the Holocaust* (New York 1988), 29–55.
5. James Young, *Writing and Rewriting the Holocaust: Narrative and the Consequences of Interpretation* (Bloomington, IN 1988), 5, 93.
6. John Bossy, *Christianity in the West, 1400–1700* (Oxford 1985), 84.
7. Poliakov, *The History of Antisemitism*, 3:28.
8. Richard Rubenstein, *After Auschwitz: Radical Theology and Contemporary Judaism* (Indianapolis 1966), 56, 71.
9. Martin Luther, *The Jews and Their Lies*, in *Luther's Works*, trans. Franklin Sherman (Philadelphia 1971), 47:167.
10. Marlis Steinert, *Hitler's War and the Germans: Public Mood and Attitude During the Second World War* (Athens, OH 1977), 1.
11. Peter Fritzsche, *Germans into Nazis* (Cambridge, MA 1998), 8, 208–9.

12. Robert Gellately, *The Gestapo and German Society: Enforcing Racial Policy, 1933–1945* (New York 1990), 16.
13. M. Muller-Claudius, *l'Antisemitisme et la fatalité allemande* (Frankfurt 1948), 34, from *Neue Zeitung* (17 May 1947).
14. Bailey, *Germans*, 174.
15. Quoted in Steigmann-Gall, *The Holy Reich*, 141.
16. Rubenstein, *After Auschwitz*, 54–6.
17. Eriksen, "Assessing the Heritage," in Eriksen and Heschel, eds., *Betrayal*, 22.
18. See the report to Josef Goebbels on membership and finances of the Churches, 3 July 1944, in Peter Matheson, ed., *The Third Reich and the Churches* (Grand Rapids 1981), 99.
19. Matheson, *The Third Reich and the Christian Churches*, 99–101.
20. Eva Fleischner, "The Crucial Importance of the Holocaust for Christians," in Harry Cargas, ed., *When God and Man Failed* (New York 1981), 29.
21. Heer, *God's First Love*, 311.
22. Waite, *The Psychopathic God*, 33.
23. John Weiss, *Ideology of Death*: *Why the Holocaust Happened in Germany* (Chicago, 1996), 276.
24. Wilhelm Niemöller, ed., *Briefe aus der Gefagenschaft Moabit* (Frankfurt 1975), Letters of 24 October, 17 and 21 November, 24 December 1937.
25. Rubenstein, *After Auschwitz*, 47.
26. Jacob Petuchowski, "The Jewish Response: From the Viewpoint of Contemporary Judaism," *Face to Face* (Fall/Winter 1977), 10; Survivor Abraham Landau reports that while working at Auschwitz he saw a Chapel building with a cross on the roof, in *Branded on My Arm and in My Soul* (unpublished manuscript).
27. Steigmann-Gall, *The Holy Reich*, 222.
28. Steigmann-Gall, *The Holy Reich*, 247–8. See also Joachim Fest, *Face of the Third Reich: Portraits of the Nazi Leadership* (London 1970), 132.
29. W.S. Allen, "Objective and Subjective Inhibitants in the German Resistance to Hitler," in Littell and Locke, eds., *The German Church Struggle and the Holocaust*, 121–3. See also David Bankier, *The Germans and the Final Solution: Public Opinion Under Nazism* (Oxford 1992), 10.
30. Hitler, *Hitler's Secret Conversations, 1941–1944* (New York 1962), 159; John Conway, "Between Cross and Swastika," in Michael Berenbaum, ed., *A Mosaic of Victims* (New York 1990), 181–2.
31. Ian Kershaw, *The "Hitler Myth"* (Oxford 1987), 2, 5, 106, 109, 112; Hitler, *Hitler's Secret Conversations*, 83–4, 158, 296, 390. Theodor Groppe, one of Hitler's generals and a Catholic, warned Pius XII that Hitler had stated: "I will crush Christianity under my heel as I would a toad." Vincent A. Lapomarda, *The Jesuits and the Third Reich* (New York 1989), 85 n43.
32. Davies, *Antisemitism and the Christian Mind*, 59.
33. Jonathan Wright, "The German Protestant Church and the Nazi Party in the Period of Seizure of Power, 1932–33," in *Studies in Church History* 14 (1977), 397; Michael Phayer, *Protestant and Catholic Women in Nazi Germany* (Detroit1990), 47; Bankier, *The Germans and the Final Solution*, 63.

34. Phayer, *Protestant and Catholic Women in Nazi Germany*, 167–8, 74–5.
35. Kershaw, *Hitler*, 88–91.
36. Dagmar Herzog, "Theology of Betrayal," *Tikkun*, May–June, 2001. <http://www.tikkun.org/magazine/index.cfm/action/tikkun/issue/tik0105/article/010553.html>.
37. Robert Ericksen and Susannah Heschel, eds., "Introduction," *Betrayal: German Churches and the Holocaust*, 1–21.
38. Steigmann-Gall, *The Holy Reich*, 148.
39. Heer, *God's First Love*, 293; Pinchas Lapide, *Three Popes and the Jews* (New York 1967), 239.
40. Quoted by Godman, *Hitler and the Vatican*, 154.
41. Rüdiger Safranski, *Martin Heidegger* (Cambridge, MA 1998), 256.
42. Ericksen and Heschel, *Betrayal.*
43. Ericksen, *Theologians Under Hitler*, 1, 26, 33, 35, 48, 55, 58, 76, 109–19, 199.
44. Bonhoeffer, *Gesämmelte Schriften*, 2:49–50.
45. Quoted in Kenneth Barnes, "Dietrich Bonhoeffer and Hitler's Persecution of the Jews," in Eriksen and Heschel, *Betrayal*, 115–16.
46. Roy Eckardt, "How German Thinkers View the Holocaust," in Cargas, *When God and Man Failed*, 207; Uriel Tal, "On Structures of Political Theology and Myth in Germany Prior to the Holocaust," in Yehuda Bauer and Nathan Rotenstreich, eds., *The Holocaust as Historical Experience* (New York 1981), 53–7; Davies, *Antisemitism and the Christian Mind*, 110–1.
47. Martin Niemöller, *Here Stand I!* published in 1937 by Willett, Clark, in New York and Chicago, 193–8.
48. Niemöller, *Here Stand I!*
49. Niemöller, *Here Stand I!* 195.
50. Nuremberg Document NG 1531.
51. See Lawrence Stokes, "The German People and the Destruction of the European Jews," *Central European History* 6 (1973), 190; Istvan Deak, "How Guilty Were the Germans?" *New York Review of Books* (31 May 1984), 37–42; and Kershaw, *Popular Opinion and Political Dissent in the Third Reich*, 255–7.
52. Leo Stein, *I Was in Hell with Niemoeller* (New York 1942), 120.
53. Karl Barth, *Theologische Existenz Heute* (Munich 1933), 24.
54. Karl Barth, "Die Kirche und die politische Frage von heute," in *Eine Schweitzer Stimme, 1938–1945* (Zurich 1945), 80.
55. Karl Barth, *Church Dogmatics* (Edinburgh 1957), 2:208–9.
56. Karl Barth, *Against the Stream* (London 1954), 198.
57. Quoted in Herzog, "Theology of Betrayal."
58. Snoek, *The Grey Book*, 291–5.
59. Niemöller, "The Niemöller Archives," 54.
60. Snoek, *The Grey Book*, 291–2.
61. Martin Niemöller, *Not und Aufgabe der Kirche in Deutschland*, in Philip Friedman, "Was There an 'Other Germany' During the Nazi Period?" *YIVO Annual of Jewish Social Science* (New York 1955), 10:82–124.
62. Gutteridge, *Open Thy Mouth for the Dumb*, 306, n7.
63. Michael Faulhaber, *Judaism, Christianity, and Germany* (London 1934), 13–14.

64. Heer, *God's First Love*, 322.
65. Yisrael Gutman and Shmuel Krakowski, *Unequal Victims* (New York 1986), 1–26; Artur Sandauer, "On the Situation of a Polish Writer of Jewish Descent," *Pisma Zebrane* (Warsaw 1985), in Abraham Brumberg, "Poland and the Jews," *Tikkun* (July/August 1987), 17.
66. Godman, *Hitler and the Vatican*, 124–5.
67. George Mosse, *Nazi Culture* (New York 1966), 256–61; Heer, *God's First Love*, 311; Lapide, *Three Popes and the Jews*, 239; Léon Papeleux, *Les Silences de Pie XII* (Brussels 1980), 77.
68. Mosse, *Nazi Culture*, 256–61; Heer, *God's First Love*, 311; Lapide, *Three Popes and the Jews*, 239; Papeleux, *Les Silences de Pie XII*, 77.
69. Gutteridge, *Open Thy Mouth for the Dumb*, 219, n17.
70. Lapide, *Three Popes and the Jews*, 239.
71. Heer, *God's First Love*, 330.
72. Gutteridge, *Open Thy Mouth for the Dumb*, 219, n17.
73. Alfred Delp, *Zur Ende Entschlossen* (Frankfurt 1949); H. Portmann, *Der Bischof von Münster* (Münster 1947); Lapomarda, *The Jesuits and the Third Reich*, chapter 1.
74. Saul Friedlander, *Kurt Gerstein* (New York 1969), 136.
75. Pierre Blet et al., eds., *Actes et Documents du Saint Siège relatifs à la Seconde Guerre Mondiale* (Vatican City 1965–75), 2:318–27.
76. Lucy Dawidowicz, *The War against the Jews* (New York 1975); Eberhard Jackel, *Hitler's Weltanschauung* (Middletown 1972), 53; Sarah Gordon, *Hitler, Germans and the "Jewish Question"* (Princeton 1984), 107.
77. Klaus Scholder, *Die Kirchen und das Dritte Reich* (Frankfurt 1977), 320–53.
78. Kershaw, *Popular Opinion*, 154.
79. "The Roosevelt–Noguès and the Roosevelt–Giraud Conversations at the President's Villa (noon and 4:20 P.M., January 17, 1943)," Roosevelt Papers, McCrea Notes, in *Foreign Relations of the United States: The Conferences at Washington 1941–1942 and Casablanca 1943* (Washington, DC 1968), 608–11.
80. Gellately, *Gestapo and German Society*, 184.
81. Pierre Vidal-Naquet, "Discourse–Memory–Truth," *Assassins of Memory*, trans. Jeffrey Mehlman (New York 1992) <http://www.anti-rev.org/textes/VidalNaquet92b/part-3.html>. See also the diary entry of SS Dr. Johann Paul Kremer for Se5, 1942.
82. Marcel Ophuls, *The Sorrow and the Pity*, interview.
83. G. van Roon, *Neuordnung im Widerstand* (Munich 1967), 561.
84. Hermann Graml et al., *The German Resistance to Hitler* (Berkeley 1970), 112.
85. In *Bonhoeffer: A Documentary* by Martin Doblmeier, quoted in Herzog, "Theology of Betrayal."
86. Weiss, *Ideology of Death*, 366.
87. See Joachim Remak, ed., *The Nazi Years: A Documentary History* (New York 1969), chapter 11.
88. Steigmann-Gall, *The Holy Reich*, 228–9.

89. Detlev J. K. Peukert, *Inside Nazi Germany: Conformity, Opposition, and Racism in Everyday Life* (New Haven, CT, 1982), 60; Gellately, *Gestapo and German Society*, 206, 207, 214. Ophuls, *The Sorrow and the Pity*.
90. Weiss, *Ideology of Death*, 369.
91. Report Submitted by Alfred Wetzel to the Ministry of Eastern Occupied Territories, "On the Organization of the Occupied Regions of the USSR." Berlin, April 1942. Nuremberg Document NG 2325. Poliakov, HARVEST, 262.
92. Pierre Ayçoberry, *The Social History of the Third Reich, 1933–1945* (New York 1999).
93. Kurt Pätzold, "Terror and Demagoguery in the Consolidation of the Fascist Dictatorship in Germany, 1933–34," in Michael Dobkowski and Isidor Wallimann, eds., *Radical Perspectives on the Rise of Fascism in Germany* (New York:1989).
94. Hitler, *Hitler's Secret Conversations*, 159; Conway, "Between Cross and Swastika," 181–2; Kershaw, *The "Hitler Myth,"* 2, 5, 106, 109, 112; Hitler, *Hitler's Secret Conversations*, 83–4, 158, 296, 390.
95. Robert Eriksen, "A Radical Minority: Resistance in the German Protestant Church," in Francis Nicosia and Lawrence Stokes, eds., *Germans Against Nazism: Nonconformity, Opposition, and Resistance in the Third Reich* (New York 1990), 119.
96. Jaroslav Pelikan, "Grundtvig's Influence," in Goldberger, *The Rescue of the Danish Jews*, 174, 176–7, 179–80.
97. Weiss, *Ideology of Death*, chapter 11.
98. Arthur May, *The Hapsburg Monarchy* (New York 1951), 179.
99. Weiss, *Ideology of Death*, 330.
100. See Herbert Ziegler, *Nazi Germany's New Aristocracy* (Princeton 1989), 83; Fleischner, "The Crucial Importance of the Holocaust for Christians," 29.
101. Quoted by Peter Padfield, *Himmler* (New York 1991), 3.
102. Weiss, *Ideology of Death*, 173, 277.
103. Weiss, *Ideology of Death*, 387.
104. Georg Glockemeier, *Zur Viener Judenfrage* (Leipzig 1936), 106.
105. Hitler, *Mein Kampf*, 6.
106. Weiss, *Ideology of Death*, 184, 191–2; Paul Johnson, *A History of the Jews* (New York 1987), 472.
107. Konrad Heiden, *Der Fuehrer: Hitler's Rise to Power* (Boston 1944, 1969), 49.
108. Hitler, *Mein Kampf*, 52.
109. Weiss, *Ideology of Death*, 163.
110. Hitler, *Mein Kampf*, 57, 324. See also Alan Bullock, *Hitler: A Study in Tyranny* (New York 1964), 40; and Elias Canetti, *Crowds and Power* (New York 1962), 47.
111. Hitler, *Mein Kamp*, chapter 3.
112. Heer, *God's First Love*, 130.
113. Mosse, *Toward the Final Solution*, 205.
114. Heer, *God's First Love*, 130.
115. Heer, *God's First Love*, 284–6.

116. Kurt Ludecke, *I Knew Hitler* (London 1938), 465–6, in *The Speeches of Adolf Hitler*, ed. Baynes, 369 n1.
117. Hitler, *Mein Kampf*, 65.
118. E. Roy Eckardt, *Your People, My People* (New York 1974), 22.
119. Bytwerk, *Julius Streicher*, 47.
120. Showalter, *Little Man*, 104–6, 213–16.
121. Bytwerk, *Julius Streicher*, 57, 62, 33.
122. Nuremberg Document PS 2699, *Der Stürmer* of Christmas 1941.
123. Eckart, *Der Bolshevismus von Moses bis Lenin: Zwiegespräch zwischen Adolf Hitler und mir* (Munich 1924), 20–1.
124. Langmuir, "Toward a Definition of Antisemitism," 309–10.
125. C. Sailer and G.J. Hetzel, *Die verfassungsfeindlichen Umtriebe der Ev.Lutherischen Kirche in Bayern*, o.J., 14, cited in Hubertus Mynarek, "Martin Luther: Psychopath oder Kriminaler?" <http://www.buerger-beobachten-kirchen.de/literaturtipps/die_neue_inquisition.html>
126. Helmut Schramm, *Der jüdische Ritualmord* (1943), in Dundes, *The Blood Libel Legend*, 348–9.
127. Nuremberg Document No. 2527, in Henry Monneray, *La Persécution des Juifs dans les pays de l'Est* (Paris 1949), 71–2.
128. Waite, *The Psychopathic God*, 35–6.
129. Waite, *The Psychopathic God*, 35–6.
130. Hitler, *My New Order*, 26–7.
131. Hitler, *Mein Kampf*, 307.
132. Quoted in Steigmann-Gall, *The Holy Reich*, 115.
133. *The Speeches of Adolf Hitler*, ed. Baynes, 240, 386–7.
134. *The Speeches of Adolf Hitler*, ed. Baynes, 370–1.
135. Eriksen, "A Radical Minority," 119.
136. Lapide, *Three Popes and the Jews*, 239; Weiss, *Ideology of Death*, 390.
137. Heer, *God's First Love*, 309.
138. Bernhard Stasiewski and Ludwig Volk, eds., *Akten deutscher Bischöfe über die Lage der Kirche 1933–1945* [6 Bde, bearb. von Bernhard Stasiewski (Bd.I–III) und Ludwig Volk (Bd.IV–VI), Mainz 1968–1985] (Mainz 1968), Nr. 32/I:101–02.); Lapide, *Three Popes and the Jews*, 90.
139. Godman, *Hitler and the Vatican*, 32, 35.
140. Kershaw, "*The Hitler Myth*," 107.
141. Bailey, *Germans*, 196.
142. Hitler, *My New Order*, 597.
143. Quoted by Stiegmann-Gall, *The Holy Reich*, 252–60.
144. Stern, *Politics of Cultural Despair*, 149.
145. Walter Langer, *The Mind of Adolf Hitler* (New York 1972), 44.
146. Michael Schwartz, "Are Christians Responsible?" *National Review* (1980), 956–7.
147. Ernst Christian Helmreich, *The German Churches Under Hitler: Background, Struggle, and Epilogue* (Detroit 1979), 138. Helmreich lists the gifts, 489–90, n27.
148. Hitler, *Mein Kampf*, 116.

149. Heer, *God's First Love*, 286, 291–3, 307, 311, 324, 477, 484–6; Jacques Nobécourt, *"Le Vicaire" et l'Historie* (Paris 1964) 342.
150. Wright, "The German Protestant Church," 397; Phayer, *Protestant and Catholic Women in Nazi Germany*, 47; Bankier, *The Germans and the Final Solution*, 63.
151. Phayer, *Protestant and Catholic Women in Nazi Germany*, 167–8.
152. Ian Kershaw, *Hitler* (London 1991), 88–91; Phayer, *Protestant and Catholic Women in Nazi Germany*, 74–5.
153. Scholder, *Die Kirchen und das Dritte Reich*, 320, 328.
154. Wright, "The German Protestant Church," 402.
155. Lapide, *Three Popes and the Jews*, 26.
156. Christopher Browning, *Ordinary Men* (New York 1992), 176, 186.
157. Steinert, *Hitler's War and the Germans*, 1.
158. Gellately, *The Gestapo and German Society*, 16.
159. Weiss, *Ideology of Death*, 276.
160. Weiss, *Ideology of Death*, 139.
161. See, for example, Nathan Stoltzfus, *Resistance of the Heart: Intermarriage and the Rosenstrasse Protest in Nazi Germany* (New Brunswick 2001).
162. Quoted by Steigmann-Gall, *The Holy Reich*, 126.
163. Eckardt, *Your People, My People*, 24.
164. Lapide, *Three Popes and the Jews*, 26.
165. Heer, *God's First Love*, 476, n276; Nuremberg Documents, "Blue Series," 12:318, 29 April 1946.
166. Hilberg, *The Destruction of the European Jews*, 12.
167. Bytwerk, *Julius Streicher*, 76.
168. Showalter, *Little Man*, 14.
169. Weiss, *Ideology of Death*, 339.
170. From Levin, *Holocaust*, 242–3, quoting Whitney Harris, *Tyranny on Trial* (Dallas 1954), 349–50; and Affidavit of Otto Ohlendorf, 5 November 1945, PS 2620.
171. Weiss, *Ideology of Death*, 340.
172. Wolfgang Gerlach, *Zwischen Kreuz und Davidstern* (Hamburg 1972), Doctoral Dissertation, 148–9, 188, 477, 488–9; Hermann Grieve, *Theologie und Ideologie* (Heidelberg 1969), 266.
173. Conway, "Protestant Missions to the Jews," 135.
174. Rubenstein, *After Auschwitz*, 74.
175. Erich Goldhagen, "Pragmatism, Function, and Belief in Nazi Antisemitism," *Midstream* (December 1972), 59–60.
176. Edward Gargan, "The Lesson of Eichmann," *Continuum* (Autumn 1966), 397.
177. Katz, *From Prejudice to Destruction*, 100–3.
178. Nobécourt, *"Le Vicaire" et l'Historie*, 176.
179. Ziegler, *Nazi Germany's New Aristocracy*, 86 n98.
180. Weiss, *Ideology of Death*, 315.
181. Ziegler, *Nazi Germany's New Aristocracy*, 92.
182. Gordon Horowitz, *In the Shadow of Death* (New York 1990), 169–70.
183. Quoted by Bullock, *Hitler: A Study in Tyranny*, 120.

184. Quoted by Frank-Lothar Kroll, *Utopie als Ideologie: Geschichtsdenken und politisches Handeln im Dritten Reich*, 2nd ed. (Paderborn 1999), 106.
185. Kroll, *Utopie als Ideologie*, 121–3.
186. Alfred Rosenberg, *Race and Race History and Other Essays* (New York 1974), 184–5.
187. Friedrich Andersen and Rudolf Homann, "The Story of Rosenberg's 'Mythus,' " *Wiener Library Bulletin*, 7 (1953), 33; Serge Lang and Ernst von Schenck, "The Story of Rosenberg's 'Mythus,' " 33.
188. Franz von Papen, *Memoirs* (London 1952), 261.
189. Hitler, *Hitler's Secret Conversations*, 400.
190. "The Story of Rosenberg's 'Mythus,' 33; Matheson, *The Third Reich and the Christian Churches*, 99–101; Friedrich Andersen and Rudolf Homann, quoted in "The Story of Rosenberg's 'Mythus,' " 33; Serge Lang and Ernst von Schenck, quoted in "The Story of Rosenberg's 'Mythus,' " 33; Rosenberg, *Race and Race History*, 68–70; von Papen, *Memoirs*, 261, quoted in "The Story of Rosenberg's 'Mythus,' " 33; Heer, *God's First Love*, 477, n291; Hitler, *Hitler's Secret Conversations*, 400.
191. Kroll, *Utopie als Ideologie*, 259, 292.
192. Steigmann-Gall, *The Holy Reich*, 131.
193. Quoted in Steigmann-Gall, *The Holy Reich*.
194. Hitler, *Hitler's Secret Conversations*, 98.
195. Steigmann-Gall, *The Holy Reich*, 131–2.
196. Steigmann-Gall, *The Holy Reich*.
197. Steigmann-Gall, *The Holy Reich*.
198. Steigmann-Gall, *The Holy Reich* 233–5.
199. Steigmann-Gall, *The Holy Reich*, 134.
200. Mosse, *Toward the Final Solution*, 119, 121, 127.
201. Christof Dipper, "The German Resistance and the Jews," *Geschichte und Gesellschaft*, 9 (1983), 372.
202. Hilberg, *The Destruction of the European Jews*, 3–4; Ezra Mendelsohn, *The Jews of East Central Europe between the World Wars* (Bloomington, IN 1983), 70–2, 118–19, 139, 141, 152, 164–6, 186, 207–8, 211.
203. Karl Schleunes, *The Twisted Road to Auschwitz* (Urbana 1970), 128–9; Hilberg, *The Destruction of the European Jews*, 43–53.
204. Leo Katcher, *Post-Mortem* (New York 1968), 231; Heer, *God's First Love*, 360.
205. Heer, *God's First Love*, 399.
206. Nobécourt, *"Le Vicaire" et l'Histoire*, 208, 358.
207. Thomas Aquinas, *De Regimine Principum et de regimine Judaeorum* (Turin 1924), 117, and *Opuscula Omnia* (Paris 1927), 1:488.
208. Aquinas, *Summa Theologiae*, 2a 2ae, 14, 3.
209. See Sterling, *Judenhass*, and Herman Greive, *Theologie und Ideologie: Katholizmus und Judentum in Deutschland und Osterreich* (Heidelberg 1969). For a secular analysis of Hitler's racism, see Burleigh and Wippermann, "Hitler's Racism," *The Racial State*, 37–43.
210. See Mosse, *Toward the Final Solution*, 119, 121, 127. "Dilettante racial theories helped give modern antisemitism the scientific veneer it lacked . . . ."

Christof, "The German Resistance and the Jews," 372. See also Rose, *Revolutionary Antisemitism*, 14.

211. For Hitler's use of the Church as a model for many elements of Nazism having nothing to do with his antisemitism, see Waite, *The Psychopathic God*, 32–6.
212. Disturbing social movements, revolutions, wars, and economic dislocations could be explained away as the results of Jewish machinations.
213. Hitler, *Mein Kampf*, 324.
214. Scholder, "Judentum und Christentum in der Ideologie und Politik des Nationalsozialismus, 1919–1945," 198.
215. Werner Masur, ed., *Hitler's Letters and Notes* (New York 1976), 210.
216. Hitler, *Mein Kampf*, 312, 316, 327, see also 305.
217. Henry Picker, ed., *Hitler's Tischgespräche im Fürerhauptquartier, 1941–1942* (Bonn 1951), 321.
218. Paul de Lagarde, *Die gegenwärtigen Aufgaben der deutschen Politik in Deutsche Schriften* (Munich 1924), 30.
219. Drumont, *La France Juive*, 2:572.
220. Stern, *The Politics of Cultural Despair*, 201.
221. In Rom. 2:28–9.
222. Aschheim, "The Myth of 'Judaization' in Germany," 230.
223. Rosenberg, *Race and Race History*, 68–70; Tal, *Christians and Jews in Germany*; and Steven Aschheim, " 'The Jew Within': The Myth of 'Judaization' in Germany," in Reinharz and Schatzberg, *The Jewish Response to German Culture*, 233–6.
224. Pulzer, *The Rise of Political Antisemitism*, 312.
225. Weiss, *Ideology of Death*, 396–7.
226. See Langmuir, "Toward a Definition of Antisemitism," 348.
227. Eckart, *Der Bolshewismus*, 46, in Aschheim, "The Myth of 'Judaization' in Germany," 240.
228. Hermann Rauschning, *Hitler Speaks* (London 1939), 229.
229. First published in French by Fayard as *Le Testament politique de Hitler: Notes Recueillies par Martin Bormann* (Paris 1959), it contains the important monologue of 13 February 1945. Republished in English as *The Testament of Adolf Hitler: The Hitler–Bormann Documents, February–April 1945* (London 1960).
230. Hitler, *The Testament of Adolf Hitler*, 55–6.
231. Berel Lang, *Act and Idea in the Nazi Genocide* (Chicago 1990), chapter 2.
232. Robert Michael, *Concise History of American Antisemitism* (Lanham, MD 2005), chapter 6.
233. Robert Wistrich argues that "Auschwitz was not built into the logic of Christianity itself but rather was the work of a heretical movement which in order to spiritually *dejudaize* Christendom aimed to make it physically *judenrein.*" *Hitler's Apocalypse: Jews and the Nazi Legacy* (New York 1985), 138.
234. Langmuir, "Toward a Definition of Antisemitism," 97. In general, Langmuir emphasizes the differences between Crusaders and Nazis, rather than the similarities.

235. The Crusader massacres have been called "the first Holocaust."
236. See Hilberg, *The Destruction of the European Jews*, esp. 4–5; Langmuir, *History, Religion, and Antisemitism*, 342.
237. Eberhard Jäckel, *Hitler's Weltanschauung* (Middletown, CT 1972), 98–9; Gordon, *Hitler, Germans, and the "Jewish Question"*; Dawidowicz, *The War against the Jews*; Milton Himmelfarb, "No Hitler, No Holocaust," *Commentary* (March 1984), 37–43.
238. Lang, *Act and Idea in the Nazi Genocide*, 192.
239. Jäckel, *Hitler's Weltanschauung*, 98–9; Gordon, *Hitler, Germans, and the "Jewish Question"*; Dawidowicz, *The War against the Jews*; Himmelfarb, "No Hitler, No Holocaust," 37–43.
240. Weiss, *Ideology of Death*, 389.
241. Tamas Nyiri, "In Lieu of a Preface," in Sandor Szenes, *Unfinished Past: Christians and Jews, Destinies* (Budapest 1986), in Asher Cohen, "Review," Holocaust and Genocide Studies 3(1) (1988), 104–6.
242. Peter Adam, *Art of the Third Reich* (New York 1992).
243. Waite, *The Psychopathic God*, 121, 139.

POSTSCRIPT

1. Peter De Rosa, *Vicars of Christ* (New York 1988), 73.
2. Herbert Vorgrimler, ed., *Commentary on the Documents of Vatican II* (New York 1969), 3:128.
3. Eva Fleischner, *Judaism in German Christian Theology* (Metuchen 1975), 146–9; Claire Huchet Bishop, *How Catholics Look at Jews* (New York 1974), 122–3; John Pawlikowski, "Jews, Judaism, and Catholic Education: Did *Nostra Aetate* Make a Difference?" in Richard Berube, ed., *After Twenty-Five Years: Jewish Christian Relations Since the Second Vatican Council's* Nostra Aetate (Symposium, Saint Michael's College, VT, 8–9 October 1990), 11; Gregory Baum, *The Jews and the Gospel* (New York 1961), 5; Davies, *Antisemitism and the Christian Mind.*

Index

CPSIA information can be obtained
at www.ICGtesting.com
Printed in the USA
LVHW020541140820
663098LV00011B/419